ETHICS and RESPONSIBILITIES of ENGINEERS

Frederick Bloetscher
Ph.D., P.E., Leed-AP

Copyright © 2022 by J. Ross Publishing

ISBN-13: 978-1-60427-184-3

Printed and bound in the U.S.A. Printed on acid-free paper.

10 9 8 7 6 5 4 3 2 1

Library of Congress Cataloging-in-Publication Data can be found in the WAV section of the J. Ross Publishing website at www.jrosspub.com/wav.

This publication contains information obtained from authentic and highly regarded sources. Reprinted material is used with permission, and sources are indicated. Reasonable effort has been made to publish reliable data and information, but the author and the publisher cannot assume responsibility for the validity of all materials or for the consequences of their use.

All rights reserved. Neither this publication nor any part thereof may be reproduced, stored in a retrieval system, or transmitted in any form or by any means, electronic, mechanical, photocopying, recording or otherwise, without the prior written permission of the publisher.

The copyright owner's consent does not extend to copying for general distribution for promotion, for creating new works, or for resale. Specific permission must be obtained from J. Ross Publishing for such purposes.

Direct all inquiries to J. Ross Publishing, Inc., 151 N. Nob Hill Rd., Suite 476, Plantation, FL 33324.

Phone: (954) 727-9333

Fax: (561) 892-0700

Web: www.jrosspub.com

CONTENTS

CHAPTER 3: States and Licensure 45

CHAPTER 4: Fundamental Education Competency 93

CHAPTER 5: Working as an Engineer 135

PREFACE

Training a new generation of engineers is a national necessity. Engineers have always been tasked with introducing evolving technologies for as long as society has existed. As a result, engineers have been agents of social changes for thousands of years. But, even while technology changes, engineers must continue to operate under a code of ethics that recognizes the obligation to protect and serve the public—placing the needs of emerging society before ourselves and our clients. This is very much the concept of *do no harm*. Just because an idea can be brought to fruition, does not mean that it should.

However, in an increasingly competitive business world, engineers are continually faced with ethical questions that balance the needs of clients with those of society as a whole. With the dynamic nature of technological growth, the ethical challenges become more and more difficult to quantify and the potential for unintended and unwanted consequences increases exponentially. The result is that conflicts with corporations, which operate under the fiduciary responsibility to protect the assets of their investors, emerge regularly. Individual profits and public service do not often align. That is why we have codes of ethics.

This book is designed to help new practitioners understand from where ethics originate and how they have developed in the profession. It is also designed to help engineers understand how the coursework they take aligns with the public good. What separates this book from others is the focus on the historical development of ethics for the profession and the role played by our educational system, accreditation commissions, and licensing boards. The knowledge and regulatory basis for the engineering occupation permits engineers to comprehend and address (and thereby often avoid) the challenges that might compromise the image of engineers in society. The trust that the public has in their judgment to protect and serve society is what allows engineers to be held in high esteem.

The political adage that *perception is reality* is true—and that is why gossip travels faster than the truth. Truth requires substantiation. As engineers, we function in an environment in which we are constrained to convey the truth in all that we do, and it is vital to maintain the positive public image of engineers that practitioners currently enjoy. It is a considerable challenge, but one that engineers are capable of meeting. Perhaps, along the way, engineers can better educate the public on the benefits of what they provide society, as opposed to being taken for granted.

ACKNOWLEDGMENTS

The author wishes to thank Charles F. Steele for his invaluable guidance, thoughts, and edits. It takes a lot for someone to volunteer their time to review a book—not once, but multiple times—and the book is far better as a result of his insight. Thank you, Dr. Steele!

ABOUT THE AUTHOR

Dr. Frederick Bloetscher is currently a professor within the Department of Civil, Environmental and Geomatics Engineering, and serves as an Associate Dean for Undergraduate Studies and Community Outreach at Florida Atlantic University (FAU) in Boca Raton, Florida. His research focus is on water resources and municipal infrastructure issues, with an interest in the sustainability of both. He received his bachelor's degree in civil engineering from the University of Cincinnati and earned his Master of Public Administration Degree from the University of North Carolina at Chapel Hill. His Ph.D. was awarded in civil engineering from the University of Miami, Coral Gables.

Dr. Bloetscher operates his own consulting firm, Public Utility Management and Planning Services, Inc. (PUMPS). PUMPS is a consulting firm dedicated to the evaluation of utility systems, needs assessments, condition assessments, strategic planning, capital improvement planning, funding options, and implementation of capital improvement construction.

Dr. Bloetscher previously served as an adjunct faculty member at the University of Miami in Coral Gables, Florida, as a former utility director and deputy director for several large municipal water and sewer systems, and served as a city manager for four years in North Carolina. He is the former Chair for the Water Resource Division Trustees, Technical and Education Council members and Education Committee for the American Water Works Association (AWWA). Dr. Bloetscher is currently the Groundwater Resource committee chair (for the fourth time), and the program and section chair for the Florida Section of AWWA. He is a Leadership in Energy and Environmental

Design-Accredited Professional (LEED-AP) and holds professional engineering licenses in multiple states.

Dr. Bloetscher co-teaches the first semester of the capstone design course at FAU, focusing on consideration of ethical issues. This course is a prerequisite to a second class where the planning and conceptual design of green building construction is turned into preliminary plans, specifications, and basis-of-design reports. Professional ethics is included as a part of both classes. Volunteering is part of his passion as well. (In the following picture, Dr. Bloetscher is shown on the left, receiving the inaugural Volunteer of the Year award from the AWWA). Dr. Bloetscher has been nominated for the Teacher of the Year award several times by his students and has received two university-wide leadership awards, as well as two national leadership awards.

In 2012, Dr. Bloetscher received the National Council of Examiners for Engineering and Surveying Award for Connecting Professional Practice and Education for his work on the Dania Beach Nanofiltration Facility, which is the first LEED-Gold water treatment facility in the world. Dr. Bloetscher was the LEED administrator for the project. Dr. Bloetscher teaches ethics classes to professional engineers in Florida as a part of their requirement for two professional development hours of ethics every two years for licensure renewal.

This book has free material available for download from the Web Added Value™ resource center at *www.jrosspub.com*

At J. Ross Publishing we are committed to providing today's professional with practical, hands-on tools that enhance the learning experience and give readers an opportunity to apply what they have learned. That is why we offer free ancillary materials available for download on this book and all participating Web Added Value™ publications. These online resources may include interactive versions of the material that appears in the book or supplemental templates, worksheets, models, plans, case studies, proposals, spreadsheets and assessment tools, among other things. Whenever you see the WAV™ symbol in any of our publications, it means bonus materials accompany the book and are available from the Web Added Value Download Resource Center at www.jrosspub.com.

Downloads for *Ethics and Responsibilities of Engineers* contain a variety of instructor materials suitable for classroom instruction.

CHAPTER 1

THE PROFESSION AND ETHICAL CONDUCT

No one willingly consults an unlicensed physician. No one submits to heart surgery conducted by an unlicensed or incompetent surgeon. Your life is in the hands of the surgeon, and your risk, if something should go wrong, is death. Because the public has very little understanding of what surgeons really know or do, they intrinsically trust what their doctors say. Few will knowingly risk their lives, and thus, most will follow the advice of their physicians. To protect the public, doctors must maintain ethical standards and follow licensing laws that regulate their profession because lives are at stake. Licensure and education are designed to indicate that the surgeon is in compliance and has met the requirements necessary to perform a surgery, prescribe medications, or assign a treatment regimen. All of these procedures and safeguards are intended to improve patient health.

Engineers are no different. Given society's continued reliance on technology and infrastructure systems for economic and societal development, the demands for new technology and upgrades to existing technology suggest that engineering is one of the most significant professions in our times. Many students pursue engineering while seeking opportunities to develop new products, design new systems, and create new infrastructure. Seeing one's design turn into a product can create immense personal satisfaction.

Industry needs engineers, and, as a result, those who pursue an engineering profession are historically rewarded with good salaries

Table 1.3 Example of subdiscipline salaries at 20 years of industry experience

Civil Disciplines	Pay at 20 Years
Construction	$119,000
Utilities	$118,000
Environmental	$109,000
Transportation	$106,000
Architectural	$101,000
Water resources	$100,000
Geotechnical	$100,000
Structural	$100,000

Source: American Society of Civil Engineers (ASCE) (2020).

power plants (Chernobyl), space vehicles (Challenger), and a host of computerized applications that are only as good as the engineers who designed them. Such events make headlines, which means the public's attention is attuned to them. That also means that the public citizenry is often very quick to identify any actions believed to be self-serving and therefore violates the public trust. This is the crossroad where the work we do and the public trust, our ethical obligation, can conflict. As a result, there is a need to understand why ethical engineering is critical, and to identify the stakeholders, audience, or evaluators who are making these judgments. This is especially true now, since the percentage of students who are scoring correctly on the Fundamentals of Engineering (FE) exam questions that specifically concern ethics has consistently declined over the past 15 years (see Figure 1.1).

To visualize why an understanding of ethics for engineers is critical, a little engineering history lesson is needed, which leads to the origins of ethics in Chapter 2, and its eventual application within the engineering occupation. Ethical questions have common threads with philosophy, thus creating the necessity for a brief discussion of how engineering ethics and philosophical concepts are interwoven. The conflicts are part of the challenge. Recognition of the value of ethical

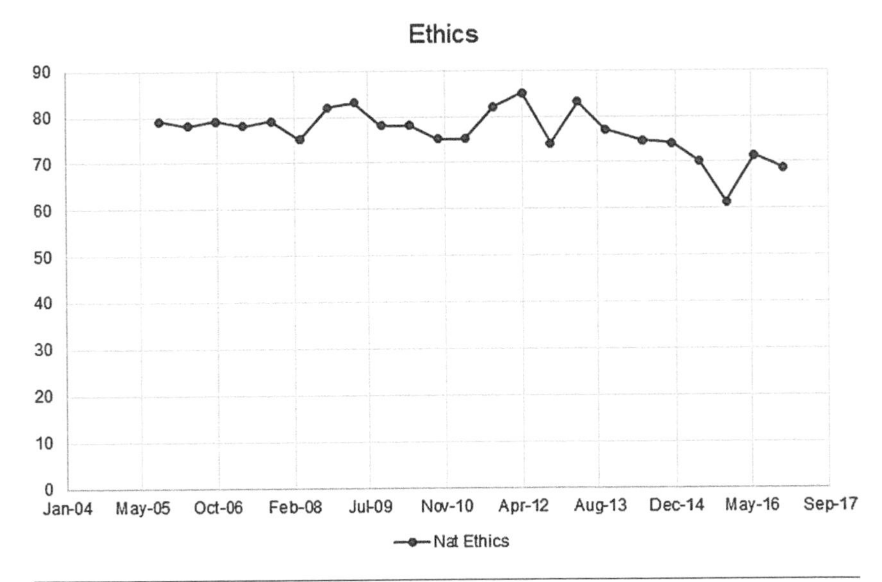

Figure 1.1 Score on the ethics questions on the FE exam from 2005 to 2017

practices has led to canons, creeds, codes, licensure, laws, education, and accreditation. Today, licensure, obtaining licensure, and the need for licensure are interwoven into the ethics discussion.

In later chapters, a discussion of education, accreditation, and vital coursework will be outlined, as will continuing education (which is required in most states). Despite the best efforts of educated and licensed engineers, ethical issues do arise—some simple, others more complex. To that end, examples of ethical challenges are presented based on cases that are generally applicable to all states.

Ultimately, the understanding of consequences and the recognition of the need for leadership are required—what we do influences and transforms society. Not all of those changes may be good. Engineers who understand the technology may be our best hope to focus things like artificial intelligence to utilitarianism for beneficial applications as opposed to predatory use. This is where the current and later chapters will take you. Enjoy the ride. If not, jump off now.

LEARNING OBJECTIVES

- Understand the role of engineers in society, and where and when ethical dilemmas may arise
- Identify characteristics of ethical practitioners
- Identify characteristics of ethical professions
- Understand how and why we differentiate between professional ethics and ethical people, and the differences between the two

1.1 WHAT ENGINEERS DO

If someone is currently pursuing an engineering career, licensure is something that may eventually be required. While certain engineering disciplines may not require licensure to the degree that civil and environmental engineers do, there are valid reasons for chemical, ocean, nuclear, industrial, transportation, aerospace, and numerous other *degreed engineers* to obtain licenses as a part of industrial or regulatory applications.

As to be discussed later in this chapter, holding a professional engineering license demonstrates to the public that the holder has obtained the requisite education, experience, and knowledge necessary to reliably make engineering decisions and judgments that protect the health, safety, and welfare of the public. Protection of the *public health, safety*, and *welfare* is a civic duty and a public trust issue that rivals the expectations of doctors, for similar reasons.

Most people do not really know what engineers do. In fact, many people, if asked what engineers do, would say that they drive trains (see Figure 1.2). In this chapter, what engineers do and where ethical issues may arise in the everyday efforts of these professionals will be explored. Also to be explored will be historical contexts within which we will identify how concern for ethics has evolved. It is hoped that this effort will provide a clear outline for continued ethical behavior by engineers as the needs of society evolve.

Figure 1.2 An engineer, but not the one we are talking about in this book

As with the availability of medical care, the public often takes for granted the delivery of water that is used during their morning shower, the functionality of sewer lines for the shower drain and toilet flushing, as well as the integrity of the roads and bridges being used to get to work, the stormwater system that drains the roadways, and the structure of the buildings that people work in. They assume their cars will get them to work, the traffic signals with operate correctly, their computers and cellphones will work as intended, and a host of appliances they depend on will function correctly. This is done with little regard or thought of the competence or foresight of the engineers who designed these systems.

Citizens routinely assume that cars, trucks, buses, trains, and airplanes have been properly designed, and thus, are safe to use. Admittedly, members of the public assume that the heating, air conditioning, air purification, power, and communication systems we rely upon are safe, reliable, and properly designed. They assume that someone with knowledge designed and tested these systems to ensure that they are safe and reliable.

The random person on the street probably does not know how a cellular phone works, how a television set works, why cable signals come in weak or strong, or how a car works. They just have an expectation that these things will work and will continue to work. All of these projects that the public takes for granted were designed by an engineer, and while corporate entities may absorb liability for faulty projects, the success of any project relies solely on the shoulders of the design engineers.

To qualify for these opportunities, these engineers must validate their competence by obtaining a license. As a result, having a professional engineering license allows a professional engineer to perform consulting, own his/her businesses, and bid for public funding—all while continuing to demonstrate his/her competence to the public.

1.1.1 Engineers in History

Engineering is described as the profession that relies upon scientific principles to design and build things that people need. While arguments can be made that ancient people engineered weapons, housing, and defenses, the real account of *engineering* on a large scale occurred within agriculture. Agriculture fundamentally changed how humans lived—gone were the days of the hunter-gatherer that moved from place to place as food was exhausted or migrated. Agriculture reduced competition in the woods with other tribes, allowed societies to be stationary, and improved fertility since food was more plentiful and consistent. But agriculture had its limits. In many regions, rainfall was not consistent throughout the year, so droughts and dry periods could be catastrophic as populations grew. Brian Fagan notes in *Elixir* that ancient civilizations grew and died with water, and as a result, the ability to design ditches to consistently irrigate crops became a critical profession in the ancient world. Ancient Egypt and Babylonia are examples of civilizations that grew and expanded based on the ability to engineer ditches to bring water long distances and to grow crops for the masses. If fate intervened and shifted the river so that the ditches could no longer function as designed (as happened in the ancient city of Ur), the community would scatter. These ditch designers were the first civil engineers, making civil engineering among the older professions. These ancient civil engineers first needed to bring water to agricultural use. Similarly, too much water was problematic, so the ditches needed to be able to remove excess levels as well. That required even more engineering.

As irrigation designs improved, ancient communities could grow more food than they needed, thereby creating the opportunity to trade goods, which meant a need for the construction of villages with businessmen who could connect with early traders. However, in order to get goods to market, roads needed to be created. Early roads were nothing more than trails, but as the need for trade increased, roads needed to be more well-constructed and problems like topography and water bodies needed to be overcome. Roads became a major piece of the ancient infrastructure systems that had to be designed (after water supply and disposal of unwanted water).

As population centers and commerce grew, the ability to go farther in order to trade goods and the quantity of goods to be traded suggested that cities that were built near water had an advantage. Moving goods via ship was, and still is, the cheapest way to transport things. Ancient seaports, starting with the Phoenicians and continuing through the English ports of the eighteenth century, provided the opportunity for goods to be traded across the globe. A historical review of the ancient world, from Babylonia to the Romans and through the Renaissance, depicts the development of new means to address water supplies, disposal, and roads. Many of those Roman roadbeds are still used today.

The industrial revolution changed things immensely. Getting products delivered at a faster pace became a priority. Seaports expanded throughout Europe as European ships sailed around the globe. But while ships can haul large loads cheaply, they were slow and ports were limited to coastal areas. The United States had this problem until the Erie Canal was built through western Pennsylvania and into Ohio which permitted access between the resource-rich Great Lakes and the East Coast. Soon it was determined that it was easier to make steel and other materials in the Great Lakes than to ship raw materials—that is why the great midwestern cities like Detroit, Cleveland, and Chicago developed. A connection to the Mississippi River was a benefit to many ports such as St. Louis and Memphis since no ocean travel was required to get to New Orleans.

Speed required the engineering of more than ports and cobbled roads; it required new technology for transportation in general, but also to address convenience and consistency. The industrial revolution spawned railroads, cars, planes, and computers, as well as the ancillary infrastructure to support them. It is why railroads and engineering are so intertwined. The railroads of the nineteenth century were far advanced from the use of horses and wagons—moving faster, with larger loads, and crossing difficult terrain. But the need to engineer better tracks, provide proper grades, and to carefully plan bridges, water stations, refueling, and repair facilities also accelerated.

When the nineteenth century drew to a close and the twentieth century began, there had been a series of significant structural failures, including some spectacular bridge failures (most notably the Ashtabula River Railroad Disaster in 1876, the Tay Bridge Disaster in 1879, and the Quebec Bridge collapse of 1907). In part, these failures were due to fundamental changes that were occurring at the time—larger, heavier engines were going faster and hauling longer loads versus the construction demands and limitations from decades earlier. Finally, the Boston molasses disaster provided a strong impetus for the establishment of professional licensing and a code of ethics in the United States.

Transportation was the infrastructure need in the nineteenth century. In the twentieth century, the development of large cities with accompanying public health challenges was the cause for a different infrastructure need—disinfection of water to reduce waterborne illness death. This required major water supply improvements in places like California and Florida in response to sizable population growth and development. After World War II, road expansion occurred—primarily the Interstate Highway System in the 1950s through the 1970s—to improve connectivity. Next was the need to address sewage after the Cuyahoga River fires throughout much of the twentieth century. The construction of airports to move goods and, ultimately, the use of computers and the internet to increase communication have been created by and improved upon by engineers. Improvements in one area required improvements in related and ancillary services—cars were great, but roads needed to be improved, bridges needed to be built, stormwater had to be dealt with, etc.

The historical development of civilization started with engineers and continues with engineers. Any review of history illustrates how society moved forward with the engineering of new tools (or weapons) by engineers. As a result, engineers are critical agents of social change. The difference between more- and less-developed areas is often the number of engineered improvements in the community. The involvement of engineers in the development of civilization creates a responsibility to the public health, safety, and welfare of the community.

Today, both public and private sector entities require goods and services, as well as the need to construct and acquire capital facilities on an ongoing basis. New technologies must be developed to provide services faster, more efficiently, and at less cost. With proper planning and consideration of societal needs, new projects and programs can be developed to benefit the local community.

Planning is required to anticipate needs simply because the existing processes that are being used to deliver these facilities require time and effort. Given that a large portion of the public does not really know what engineers do, yet expects that the job is done correctly while protecting everyone's interests, engineers need to communicate what they do and also make good decisions. This has become more important than ever, given that the infrastructure that built this nation is now crumbling around us—as noted in the infrastructure grades that are handed out by the ASCE every four years (see Table 1.4).

Table 1.4 ASCE infrastructure grades

Infrastructure Category	2001 Grade	2005 Grade	2009 Grade	2013 Grade	2017 Grade	2021 Grade
Aviation	D	D+	D	D	D	D+
Bridges	C	C	C	C+	C+	C
Dams	D	D	D	D	D	D
Drinking water	D	D–	D–	D	D	C–
Energy (national power grid)	D+	D	D+	D+	D+	C–
Hazardous waste	D+	D	D	D	D	D+
Inland navigable waterways	D+	D–	D–	D–	D–	D+
Levees	—	—	D–	D–	D–	D
Ports	—	—	—	C	C	B–
Public parks and recreation	—	C–	C–	C–	C–	D+
Rail	—	C–	C–	C+	C+	B
Roads	D+	D	D–	D	D	D
Schools	D–	D	D	D	D	D+
Solid waste	C+	C+	C+	B–	B–	C+
Transit	C–	D+	D	D	D	D–
Wastewater	D	D–	D–	D	D	D+
Overall	D+	D	D	D+	D+	D+

Given that engineers require public trust, students and practitioners need to be able to immediately identify anything that may raise ethical questions in the engineering field. Remember, most people do not understand what engineers do, they just expect that the engineers do it right and that the things they rely on (or paid for) will work properly. That simple trust suggests the requirement to frame a concept of ethics. But what are ethics? It seems like something to do with public expectations of competence and an application of judgments which are considered good—but that leaves us with only a foggy idea of a perception of ethics and no real answers. Let us see if the fog can be cleared.

1.2 ENGINEERING ETHICS

Ethics is an issue that surfaces in the engineering world on an ongoing basis (some states even require a formalized ethics refresher course at defined intervals of time). But what are ethics? To begin to answer this question, we must start with philosophy. A cursory review indicates that there are three potential definitions of a person with ethical behavior (Popkin and Stroll, 1993):

- One who establishes a set of values and lives by them
- One who lives by any set of values which is shared by a group of people
- One who lives by a set of values that is universally accepted

Let us look at each one of these. The first definition *is a person who establishes a set of values and lives by them.* What does one make of this definition? Is it acceptable? Do we accept a person who acts this way? In reality, few people accept this first definition of an *ethical person* because values can vary and may include individuals with a highly personalized set of ideals (e.g., Robin Hood) or individuals with frequently unacceptable behaviors (e.g., serial killers).

So obviously, *a person who lives by any set of values which is shared by a group of people* must be better. What about this definition? Is it acceptable? Do we accept a person who acts this way? These people share many of the same beliefs and conform to an accepted set of *rules* of acceptable behavior. Engineers are among groups with *common values.* That sounds good, but alas, there are many such groups including religious cults and political parties purporting ethical behavior with which we may not necessarily and fundamentally agree with. Worse still, groups with common *values* include terrorists, fascists, racists, white supremacists, neo-Nazis, and many others whom we do not support. So, this definition really does not work either.

Therefore, the third option—could *a person with a set of values that is universally accepted* be the answer? What does one make of this definition? Is it acceptable? Do we accept a person who acts this way?

Find *one* example of a universally accepted ethical value. Just one! A common one is *do not kill*, but what does that make of members of our military? Another is to *always be honest*, a worthy and perhaps the closest thing to a universal ethic. The truth can hurt and most people do not want to create bad feelings, so honesty, while being the best policy, does have its limitations. So, these specific philosophical answers are not very helpful in defining ethics for the engineering profession.

Another approach is to examine the ethical systems within individual professions. What professions do most people perceive to be ethical or unethical? (Ignore for the moment whether the perception is reality or not.) Professions that are perceived to be unethical by the public on a routine basis include:

- Salesmen of any type
- Lawyers of any type
- Politicians of any type
- Financial brokers and bankers
- Realtors
- Mechanics
- Contractors

Sometimes illegal enterprises are included, but illegality is not necessarily germane as an ethical consideration. Organized crime typically operates from a set of ethical values and core principles that are sworn to. That does not make the actions of these organizations acceptable to society by any stretch of the imagination, but there is a fundamental set of *ethics* within these organizations that fits any working definition.

In a classroom setting, prostitution is often mentioned as an *unethical activity*, but if you get what you pay for, certain economists will ask what is unethical? (Let us set aside for a moment the issues associated with forced labor, human trafficking, the view of women in general, and other clearly unethical activities that may be included in the profession.) The fact is that prostitution is illegal (except in Nevada) and perhaps immoral, but unethical? That is not as clear; and we could say the same when it comes to several criminal activities—including the illegal distribution of drugs.

Ultimately, the commonality among professions that is perceived to be unethical is the fact that a *transaction* is associated with the movement of money. Those who are perceived as being unethical must receive a transaction of money. A contract, by definition, must include consideration on both sides. So, let us compare these professionals (or occupations) to professions that are generally perceived as being ethical:

- Engineers
- Scientists
- Medical personnel
- Teachers
- Public safety workers
- Healthcare providers
- Social workers

The perception is that people who work in these professions serve the public and protect the public interest. Keep in mind that many of these professions (and a few others) involve people to be relied upon as a part of the overall efforts to contain the coronavirus pandemic. The public has expectations that those who are practicing in these professions know what they are doing even though the public may not understand what they actually do. There is a trust factor associated with these perceived-to-be ethical professions that are expected to combine competency with accountability. As a result, most practitioners within these fields must be licensed, which means they are regulated. If things go wrong with one of these service providers, they can be brought before regulatory boards and reprimanded, disciplined, or terminated.

Another common trait among those professions that are perceived to be highly ethical is that practitioners find most of their decisions are based on human judgment. Doctors, medical staff, and first responders, similar to engineers, have imperfect information, but they make a diagnosis based on their *best guess*, given the facts that are available.

However, the public expects that these practitioners will provide the correct assessment every time. This is the same with teachers who are trying to reach all of their students when those students are all obviously unique and respond to different stimuli. Many of these situations will not provide definitive answers, and furthermore, these situations are always dynamic and in flux. The concept of licensure stems from responsibility to the public and the expectation of the public that engineers will act to protect their interests.

1.3 SUMMARY

So, what should be learned from this chapter? First is that engineers have a major role in society and that virtually everything citizens use in their daily lives involves some form of engineering. The fact that engineered products are ubiquitously around us may lead to a diminishing perception of the importance of engineers to society. Engineers clearly need better marketing.

The next step is to review ethics. It is easier to identify unethical behavior than define it. It is not possible to find a set of universal ethics, but social or public expectations go a long way toward defining what is and is not ethical—we see that in the perception of unethical professions. This chapter ends with the realization that it is the public expectations of competence that differentiate ethical from unethical professions, but that there is no clear answer.

REFERENCES

ASCE 2020 New Report: Civil Engineers' Income and Job Satisfaction Continue to Rise. https://www.asce.org/templates/press-release-detail.aspx?id=38963#:~:text=reston%2c%20va.,from%20%24109%2c000%20reported%20last%20year.

Bloetscher, F. and D.M. Meeroff. (2015). *Practical Concepts for Capstone Design Engineering.* J. Ross Publishing. Plantation, FL.
Michigan Tech. (2020). 2020 Engineering Salary Statistics. https://www.mtu.edu/engineering/outreach/welcome/salary/.

PROBLEMS

1. Obtain the latest information available for local, regional, and national salary trends for your field of engineering. How do these trends compare to other professions?
2. Develop a table to help make these comparisons easier to understand.
3. Name and describe a universally accepted value.
4. Among the common answers to the question "what do engineers do?" is:
    a. Drive trains
    b. Get into politics
    c. Go to the Moon
    d. Act like Sheldon in the Big Bang Theory
5. Professions that are often deemed to be ethical include:
    a. Doctors
    b. Lawyers
    c. Politicians
    d. The mafia
6. Illegal activities are always unethical.
    a. True
    b. False
7. The priority for engineers is to:
    a. The public
    b. Their client
    c. The firm's shareholders
    d. The mayor

8. An ethical person is a person with a set of values and lives by them.
    a. True
    b. False
9. The expectation of the public is that engineers will:
    a. Design buildings to protect the public
    b. Report potential failures to their clients only
    c. Will tackle any problem handled to them
    d. Will find definitive answers
10. Judgment is required by engineers because:
    a. There is often imperfect information
    b. It is better than guessing
    c. It allows them to find the perfect answer to the problem
    d. There is often imperfect education

This book has free material available for download from the Web Added Value™ resource center at *www.jrosspub.com*

CHAPTER 2

FROM WHERE DO ETHICS EVOLVE?

In Chapter 1, we attempted to determine what identifies ethical people. However, it was easier to find professions that were deemed to be unethical and ones that appeared to be more ethical, but it was less clear as to why. In this chapter, we will take a look back in time to see if we can find the origins of when the concept of ethics formed. This chapter will build upon the history from Chapter 1, with more focus on how professions have developed over the past 10,000 years or so and why. Greek philosophers weighed in on the subject of ethics 2,500 years ago, but earlier rulers had their own ideas.

Since the time of the Greeks, the study of ethics entered a period of several thousand years of development and refinement, especially as it applies to professions. By the nineteenth century, professional societies were being specifically chartered for engineers. The professional societies led directly to state regulatory bodies as long as 110 years ago. Our focus will now turn to what the professional societies present concerning engineers and ethical behavior. Let us sit back and have a little history lesson.

LEARNING OBJECTIVES

- Gain an understanding of a brief history of ethics within the engineering profession
- Understand why and how ethics developed
- Understand how ethics evolved for various professionals
- Understand how ethics developed for the engineering industry today

2.1 A BRIEF HISTORY OF THE DEVELOPMENT OF ETHICS

Around 10,000 B.C., human civilization consisted of disparate bands and tribes, basically hunter-gatherers that subsisted on wild animals and plants. The person who brought food home was respected. The elders were respected for their knowledge as well as their leadership. Shelter providers and those loyal to the group also ranked highly.

Members who were considered food hoarders, people who damaged member property, selfish individuals, thieves, or freeloaders were not appreciated. Offenders were, in many cases, cast out and banished from the group, condemned to survive alone in the wild. At that time, few people, if any, were capable of survival alone in the wilderness, so this banishment was somewhat of a quasi-death sentence. The tribe, in order to be successful, figured out how to control its members to assure the survival of the collective.

Legends were passed down to demonstrate ethics and propagate acceptable behavior. Many current fairy tales include references to the *scary forest* (Snow White comes to mind). Welcome to ancient Druidic lore, where being cast out into the woods was the scariest penalty of all. Those tribes that were successful developed by focusing on the collective good; we know little of those who did not.

Somewhere around 7,000 B.C., humans began to master agriculture, which fundamentally altered the history of civilization and the world. No longer were humans dependent on grazing the land for animals and berries—nor did they need to move because the food ran out. With the cultivation of grains and certain animals, humans could stay in one place and continue to grow new crops and raise animals for food, which permanently changed the way people lived. Staying in one place meant that people could put down roots and develop communities.

Agriculture meant better nutrition, which meant larger populations could be supported, which in turn, meant a need for more agriculture. With more harvests came the opportunity of a surplus, and efficiency that led to economic benefits via trade with other communities. Since agriculture is stationary, there needed to be a marketplace to trade goods. Since people settled in one place for agriculture, markets and villages were established for trading purposes.

To increase agricultural efficiency in order to provide more productive harvests, it became necessary to manage water supplies for irrigation, which was one of the earliest of engineering feats. Efficient water engineering resulted in a surplus of food, allowing others in the community to become free to specialize in other tasks to support the community. While the farmers specialized in growing wheat, barley, or grapes, for example, others specialized in animal husbandry, making tools, processing food, healing the sick, building defenses, and trading with other tribes.

The villages also needed to build infrastructure to maintain order and promote efficiency. This infrastructure included shelter to protect families from the elements, defenses to protect the village from incursions, roads to transport goods, and storage facilities to protect the harvest. They also needed places to do business and they needed funds to pay for all of these facilities. Welcome to taxation.

Thriving groups of villages became city-states that traded on a larger scale. They needed more roads, bridges, ports, more complex irrigation and drainage systems, and methods to deal with wastes—all early engineering projects. Since many of these projects were community-based, taxes and commerce were developed. Laws to regulate trade among people who were unknown to each other were also created to ensure fair dealing. Laws were later recorded, and trade centers with specialties were created.

The best specimens of the products were in high demand, but the producers were dependent on others to provide packaging and shipping. This dependency created expectations that the packagers knew what they were doing—again, trust was essential. Likewise, the shippers were trusted to know the routes and to provide reliable transportation equipment. As demand outstripped a given merchant's ability to supply, more merchants were needed, but the expectation was that the quality would remain. Products that failed could be catastrophic for the village, so training guilds and similar orders were created to ensure that the workers provided the level of expertise and workmanship that the public expected in a free and competitive environment. Thus, the public expectation that people would do their jobs correctly was beginning to become a behavioral ethic.

EXAMPLE 2.1

Passing Knowledge Down

Let us assume that your community has developed excellent agricultural processes that are used to make a high-quality beer. The beer is of such high quality that it is in high demand in other communities that are willing to trade for it. Since your community has invested heavily in developing the beer and is highly successful at making it, the village soon becomes very wealthy. The ultimate law of the land prohibits the spillage of the beer since beer is the main source of income for the community. Ultimately, an issue arises whereby the barrel maker is no longer able to keep up with the supply of beer. So, a new vendor, we will call him Bob, starts making barrels. However, Bob's barrels leak, which is a violation of the laws against alcohol abuse and spillage. What should be done now?

One obvious consequence would be to penalize Bob in some manner, but a smart leader would realize that such a response would not resolve the larger problem—needing more barrels to sell more beer that can be taxed to provide more infrastructure and community wealth. Thus, a better solution would be to have each of Bob's workers spend time as an apprentice under the current barrel maker. In that manner, the knowledge would be passed down, Bob would make better barrels, and the society would become wealthier by selling more beer.

This system remains much the same today—it is called an apprenticeship. In fact, this is what happened with carpentry, stone masonry, and a host of other professions along with engineering. Apprenticeships help younger people hone their skills so they can evolve and develop to be future craftsmen. In reality, the four years that engineers

continued

spend in the field working under a licensed engineer after graduation is their apprenticeship. But none of this quite gets us to the definition of ethics. That is where the philosophers weighed in. Let us settle into the past.

2.2 THE PHILOSOPHERS' PONTIFICATE

Most ancient philosophers lived in challenging times. As the Greeks developed the laws of mathematics, Greek philosophers tried to discover complementary behavioral laws to explain human behavior. The belief at the time was that any of society's failures were caused by the inability to follow some behavioral law. The ancient philosophers, starting with Plato and Aristotle, thought that education and leading a proper life were important to avoiding difficulty—so learning was encouraged. The result was philosophy, which is defined as *the study of consequences for implementation of a series of behavioral principles*. Ethics are among those behavioral principles. Answering questions about moral ideas, character, policies, and relationships is part of the process.

In the ensuing 2,000 years, philosophers—ranging from Epicurus to St. Francis of Assisi to Kant (1785), Bentham (1789), Mill (1843, 1863), and Mackie (1977)—offered their views, each contributing a small part. This resulted in three major schools of philosophical thought, each with benefits and challenges, with none answering all of life's questions, but each contributing in some way to the future and to the field of engineering:

1. Utilitarian ethics
2. Deontological ethics
3. Virtue ethics

Utilitarianism, sometimes called *consequentialism*, stems from eighteenth- and nineteenth-century English philosophers and economists, such as Jeremy Bentham and John Stuart Mill, who defined an action as being ethical if it would benefit the greatest number of people (Mill,

1863). By defining happiness as balancing pleasure over pain, they were repeating some of the same hedonistic mantras of Aristippus of Cyrene, who supported immediate gratification and pleasure and of Epicurus, who suggested that pleasure coincides with virtue (Stanford, 2004). Today, normative hedonism is the idea that *pleasure should be people's primary motivation.*

The concept of maximizing the happiness of the most people has appeal and is used in many fields. Engineers rely on alternative analyses to determine an optimal approach. In this manner, the preferred option is identified by using an objective set of selection criteria established *a priori*, while ensuring that the best-ranked alternative does not create negative consequences toward society. Engineers require accurate data to do this, as well as a reliance on observable results to meet the definition of utilitarianist principles today. However, in many cases, the outcomes are not clear ahead of the decision.

For example, utilitarianism is practiced by most democratic governments today and is supported as the basis of a free market by economists (Smith, 1776; Friedman, 1982). But the Founding Fathers in the United States realized that there could be very negative consequences to certain decisions. They created the court system in order to permit the aggrieved to appeal majority-rule decisions that created unintended consequences or that curtailed certain unalienable rights. Much of the discussion in the field of economics surrounds the conflict between completely free markets and the need to control markets for certain types of necessary goods. Otherwise only those who could afford goods could have access. Some extreme examples of goods to which everyone needs access include water and quality food. Health care is an ongoing challenge—the wealthy can afford the best care, while those who cannot, suffer as a result. Utilitarianism does not appear to resolve issues where many considerations are intuitive.

G. E. Moore (1903) was known for applying *common sense concepts* to the investigation of the effects of particular actions to determine whether an action was ethical or not (as opposed to applying intuition). In Moore's view, intuitions should reveal not the rightness or wrongness of specific actions, but whether the result was good. More recently, Anscombe (1958) expanded upon this philosophy by creating the term *consequentialism*. She believed that ethical decisions should be made based on the *consequences of the action*. The consequences of any particular action would form the basis for any valid moral judgment.

To create a structure for judgment, there are some questions that many consequentialist theories address:

- What qualifies as good consequences? (Note that we may not all agree to the definition of *good*.)
- Who is the primary beneficiary of moral action?
- How are the consequences judged and who judges them?

Barbour (2010) notes that any examination of modern ethical theories tend to focus on consequentialism and utilitarianism, both of which hold that providing the greatest benefit for the greatest number is the supreme principle. The connotation is that of societal or social benefits—something that engineers ascribe to when they protect the public health, safety, and welfare. It is noteworthy that in many cases, maintaining specific rights and obligations is a far greater requirement for engineers than just following a utilitarian approach.

There are some challenges associated with fully accepting the premises of utilitarianism, despite the perception of some potentially positive outcomes. The first is that the outcomes can be influenced by those creating the options. Part and parcel to this challenge is that rarely is a potential outcome known to the extent that "what is ethical" can be predetermined. Bowen (2009) notes that it is often impossible to accurately predict the outcomes of an action. Recognition of unintended consequences reinforces the concern that utilitarian decision making might be shallow. Utilitarianism also creates a challenge since different segments of society who might benefit from any decision might be considered more important than other segments who are less valued.

While the utilitarian philosophy is commonly seen in the study of democratic societies, the deontological approach emphasizes rights and obligations. As a result, *deontology* focuses on whether the actions are right or wrong, not the outcomes. Immanuel Kant was the father of what has now become deontological ethics. Kant held that the right thing to do is defined as *duty*—as opposed to utilitarianism that weighs social benefits. Kant was known for developing the *categorical (dialectic) imperative*, which looks for transcendent principles that apply to all humans. Kant (1785) held that "human beings should be treated with dignity and respect because they have rights." The core concept is that there are objective obligations, or duties, that are required of all people. Bowen (2009) suggests that "deontology is based on the moral autonomy of the individual." Fitzpatrick and Gauthier (2001) advocate for professionals to have a basis on which to judge the rightness of the decisions they might make every day.

Kant contributed that the concept of *duty* and the need to evaluate whether an action was ethical should be based on its impact upon society, which remains integral to engineering ethics today. The connection to engineering actions, and the reasoning for the same, are directly applicable to the engineering profession as well.

Kant proposed that people have a duty to act in a certain manner and that every action could be evaluated by a universally accepted code of behavior. Society reflects this today. The question Kant raised was, "If everybody did [insert specific activity], would society function [yes or no]?" (Mantell, 1964). If the answer was no, the behavior was deemed unethical. If the answer was yes, the practice was considered ethical. This imperative relates directly to an engineer's responsibility to protect the health, safety, and welfare of the public and, therefore, has direct applicability to ethical systems practiced within the engineering profession.

Deontology considers all humans as equal, not in the physical, social, or economic sense, but equal before "God." As a result, the concepts of equality, tolerance, diversity, and inclusion are concepts that are held within deontological ethics. W. D. Ross (1987) proposed a pluralistic form of deontology that contained seven prima facie duties: beneficence, non-malfeasance, justice, self-improvement, reparation, gratitude, and promise-keeping. All these duties are in keeping with

Kant's categorical imperative. Kant's ethics are perhaps the most extensive and consistent of the ethical mantras available.

Related are normative ethics, which consider how one might act, morally speaking, considering the rightness and wrongness of his/her actions, much as defined within Kant's philosophy.

Applied ethics is an extension whereby attempts are made to apply ethical theory to real-life situations. However, strict, principle-based ethical approaches often produce solutions to specific problems that are not universally acceptable or may be impossible to implement.

The concepts of justice, equity, and freedom are all part of the large-scale question of a deontological approach (as opposed to maintaining a narrow cost-benefit analysis that utilitarian economists might utilize). For example, when seeking to maximize social welfare, an anthropogenic goal is to develop a means to create the greatest common good. Left unanswered is how to quantify the greatest common good when the cost-benefit only focuses on the benefits while ignoring a large portion of the disadvantages or the costs. As a result of the consequences, the question that should be asked is, "What could possibly go wrong?" Ultimately, because engineers are among participating agents of social change, as potential challenges arise, engineers need to be able to identify and evaluate the magnitude of these challenges, given that engineers have a duty and an obligation to protect the public health, safety, and welfare.

While Kant's ethical theories may appear to be more complete than others, challenges to this perspective arise when there is no common agreement about the principles involved—nor is there guidance on what to do concerning negative consequences. Deontology does not provide guidance when duties conflict—there are no priorities. Anscombe (1958), MacIntyre (1985), and Brewer (2009) argue that to define categories of rightness and wrongness is already starting off on the wrong foot.

While Anscombe (1958) argued that "consequentialist and deontological ethics were only feasible as universal theories when the two schools grounded themselves in divine law that led to universal standards or ground their theories in religious conviction," a third theory known as *virtue ethics* emerged. Virtue ethics emphasized moral character, in contrast to deontology, a system that emphasizes the consequences of actions (or consequentialism) (Hursthouse et al., 2016). Interestingly, virtual ethical considerations arise from Immanuel Kant (1785a).

Virtue ethics are defined by values and behaviors that will allow a person to experience happiness and well-being (Kraut, 1989). Virtue ethics are said to mirror the ethics of Socrates and Aristotle. For example, the highest virtue was identified as inward-looking knowledge of self. Self-knowledge, considered necessary for success, was inherently seen as an essential good. Socrates noted that people will naturally do what is good, provided that they know what is right. Bad actions are seen as purely the result of ignorance.

Aristotle proposed that when a person achieves self-realization, he or she will act in accordance with virtue, do good, and be content. He also created an ethical system based on the virtues of a person when making a decision. The Indian philosopher Valluvar describes virtue as a way of life rather than any spiritual observance. Basically, it is believed to be a way of harmonious living that will lead to universal happiness. What distinguishes virtue ethics from consequentialism or deontology is the centrality of virtue within the theory (Watson, 1990; Kawall, 2009). Hume (1658) put forward a similar view on the difference between facts and values.

More recently Athanassoulis (2000) suggested that most people can truly be described as virtuous, but still have blind spots. Zagzebski (1999, 2004) likewise defines right and wrong actions by reference to the emotions, motives, and dispositions of virtuous and vicious agents. By contrast, Swanton (2003) begins with existing conceptions of virtues. Part of virtue ethics is Meta-ethics, the branch of philosophical ethics that probes how people understand and define right and wrong.

Several subsets of virtue ethics exist. An obvious one is *altruism*, which states that individuals have a moral obligation to help, serve, or benefit others, and if necessary, at the sacrifice of self-interest.

Friedrich Nietzsche criticized altruism's suggestion that the concept of being more virtuous to others than to oneself is degrading and demeaning to the self and hinders the individual's pursuit of self-development. Rand (1964) stated that most problems in the world come from the doctrine of altruism because "there is no rationale for asserting that sacrificing yourself in order to serve others is morally superior to pursuing your own self-interest."

Naturalistic ethics suggest that there are objective moral properties to which people have empirical knowledge, but that these properties are reducible to entirely natural properties (such as needs, wants, or pleasures). These properties sound more like hedonism than virtue ethics unless pleasure is the virtue being sought. Epictetus said the greatest goals were contentment and serenity, along with mastery over one's desires and emotions, which seems virtuous in a world laden with much vitriol.

Sustainability has been added as a value of engineering ethics over the past 20 years. Sustainability focuses on the protection of resources for future generations. This is not incongruent with environmental ethics, which also considers the relationship between humanity and the natural environment. Sustainability addresses ways that maintain and protect natural resources such as water, air, soil, wetlands, and forests (https://ethicsunwrapped.utexas.edu/glossary/sustainability).

Environmental ethicists do not believe in transforming nature for the benefit of mankind. To further this vision, engineers must increase awareness of the need to make the world fairer and happier (bringing in principles of all three previously presented systems to ensure that the needs of future generations will be met). For this reason, the engineer should avoid any action that is intended to harm the general interest, thus avoiding a situation that might be hazardous or threatening to the environment, life, health, or other rights of persons. This concept has been added to engineering codes of ethics over the past 20 years.

However, virtue ethics has limitations since it tends to miss the importance and obligations to the public and the clients. There are also difficulties in the application of virtue ethics, including conflicts between self, competing goals, and clear justification—in many cases, virtue ethics exist in the territory where much is unknown, like the breadth of consequences for the failure to adhere to the sustainability mantra.

Compounding the challenge are societal norms that can change with time. As a result, professions and applicable systems of ethics may need to change rapidly to keep up with societal norms. An example is the integration of artificial intelligence into society. The concept has great potential to make things more efficient, but there are many serious ethical challenges, as will be discussed in Chapters 3 and 7. Perry (1992) notes that postmodern ethics endorse a greater acceptance of the view that people can do whatever they wish as long as it does not adversely affect others (which is not completely in contrast to the ethics provided by earlier philosophers). This attitude may be very self-serving but does reflect attitudes that are being demonstrated today. The Covid-19 pandemic is a prime example in which people have refused to comply with mandates, often because the mandates interfered with what they might want to do or inconvenienced them in some way (whether real or perceived).

All three moral theories provide significant guidance for certain questions but are limited when universally applied. The following extract compares the basics of each ethical theory. Skeptics suggest that all three theories share the same challenges (Stanford University, 2002):

> "(a) *no one is ever justified in believing that moral claims (claims of the form "state of affairs* x *is good," "action* y *is morally obligatory," etc.) are true and, even more so*
>
> (b) *no one will ever know that any moral claim is true. Moral error theory holds that we do not know that any moral claim is true because:*
>
> (i) *all moral claims are false,*
>
> (ii) *we have reason to believe that all moral claims are false, and*

> *(iii) since we are not justified in believing any claim, we have reason to deny, we are not justified in believing any moral claims."*

That does not make the consideration of ethical systems inappropriate for engineers. In fact, engineers should take the best of each system in application to the profession. It means engineers need a system to apply ethical or moral reasoning to every situation they may face. Ethical reasoning is a part of ethical understanding. It requires that the person must consider *moral reasonableness.* That brings the concept of *social justice* ethics to the fore. Social justice theorists worry about *distributive justice*—that is, what is the fair way to distribute goods among an inclusive group of people? For engineers, concern for the location of a facility may have an adverse impact upon at-risk communities. This is part of the concept of the social contract that Socrates suggested 2,500 years ago. Modern social contract theorists, such as Thomas Donaldson and Thomas Dunfee (*Ties that Bind*, 1999), observe that various communities make rules for the common good. The rights that people have—in positive law—come from whatever social contract exists in the society.

Moral reasoning also requires a respect for people (as does Kant), reflected in a general concern for well-being. This is also part of the social contract. Hope, tolerance, and integrity round out the expectations of ethical reasoning skills. When in place, Thomas Hobbes believed that people in a *state of nature* would rationally choose to have some form of government. He called this the social contract, where people give up certain rights to government in exchange for security and common benefits. This view differs from that of the deontologists and that of the natural-law thinkers who believe that rights come from a divine being or some transcendent moral order.

The social contract extends to recognition of engineering as a profession by the public. While most people assume that *profession* means an occupation, the term extends beyond just a job. Professions are deemed to require advanced or specialized expertise that is not commonly available in the public realm. It also assumes that professionals will continue to grow and develop in their skill set to meet new and changing demands and technologies. Professionals are assumed to meet the tests of public interest, whereby their actions are focused on the public good.

All of this requires some moral or ethical standard by which people can deal with professionals. An important justification for a code of ethics (COE) applied to professional engineers is that it sets a standard for professional behavior.

Furthermore, since the public generally does not collectively have the knowledge to know or understand what engineers do, there is some need to self-regulate the profession—in order to provide a protection that will keep unqualified people from offering services that potentially cannot provide the same degree of protection concerning the public health, safety, and welfare. Self-regulation, discussed in Section 2.3, starts with published creeds and codes of ethics that the public can understand and interpret. The ethical concepts of honesty, integrity, respect, and tolerance are part of the public expectation.

Schwartz (2016) notes that ethical systems that were developed over the centuries by philosophers may not provide answers or solutions to the ethical dilemmas that are currently faced by engineers, but they might provide guideposts that can be helpful in assisting engineers in evaluating the circumstances they may face. Each ethical system serves as a good reference point for engineers who seek to ethically meet the goals and needs of the clients (while at the same time protecting the public and designing sustainability into the future).

2.3 CREEDS, CODES, AND CANONS

As engineering developed further into the nineteenth century, it was inevitable that various accidents would occur. The U.S. Military Academy at West Point was the first engineering school to be established in the United States (1819). Others followed thereafter, but a means to communicate among those within the profession was lacking.

The first professional engineering society was the Institute of Civil Engineers in Great Britain in 1828 (American Society of Civil Engineers (ASCE) website). Various attempts were made to develop a similar society in the United States in the 1830s, culminating with the first engineering society in the United States, the Boston Society of Civil Engineers in 1848—shortly thereafter succeeded by ASCE in 1852. Various organizations followed, including the National Society of Professional Engineers (NSPE) in 1934 after the majority of states in the United States had created professional engineering licensing boards. Each developed an individualized set of guidelines for engineers—all dubbed codes of ethics—which often included a series of creeds and canons.

As the engineering profession developed specific rules, it began to self-regulate and create disciplinary actions for failure to adhere to its rules. There is a theme that underlies all their guidelines today: *the top priority under ALL CIRCUMSTANCES is the health, safety, and welfare of the public.* Legal issues are second, and third is the engineering profession itself. This assured that engineers would be perceived to be ethical, and emphasized the perception that the incompetence and liability of one practitioner would eventually damage the reputation of all engineers and the profession. The client is fourth on the priority list, and the engineer is last on the hierarchy for this reason.

There are a series of other provisions to which engineers are typically required to abide. The first is that engineers must perform work only within their area of expertise, and not in any area where they may lack experience or competence. Engineers must not accept a project if they do not have the skills to design it. In other words, they must recognize their competencies as well as their personal limitations. This is directly related to the protection of the public health, safety, and welfare.

Closely related to protecting the integrity of the profession, engineers are required to issue objective and truthful comments and avoid conflicts of interest. Engineers must build their personal reputation on *merit*, by upholding the dignity of the profession and continuing to pursue professional development throughout their careers.

Due diligence is a legal concept that involves ensuring that all information required to make a decision is properly considered. Performing due diligence will be necessary for whomever is ultimately responsible (and for whomever is funding the project) to write an accurate report. When engineers hear the words *due diligence*, it is likely from a lawyer that is arguing that the parties involved neglected it.

In protecting both the public health and the professions, the canons state that no one is qualified to review the plans of a professional engineer unless they are also a professional engineer. This ensures professionalism in the process. With that said, engineers may be paid by someone to review another engineer's plans, but the involved engineer must be notified in writing. One reason for doing this is that all of the information is probably not within the plans and specifications, so that needed information is provided to achieve appropriate judgment or review. Engineers need to practice due diligence when reviewing someone else's work.

There is a need to protect the client (priority item #4). Engineers are only to be paid by their clients. In fact, the client's fee must be their *only source* of income. Engineers cannot accept money from suppliers, competing clients, manufacturers, or salesmen, among others, because doing so would compromise their ability to enforce the requirements within the plans and specifications. Therefore, accepting compensation from someone other than the client, on any project, is a conflict of interest that engineers must avoid. ASCE goes on to note that engineers should protect the client's interests, as long as doing so does not conflict with protecting the public interest.

If things go wrong, engineers are required to freely and openly admit errors, and then offer a solution. If engineers have no involvement in an issue, they should not comment (a legal concept called *standing*). To help this process, the engineer's creed was developed.

NSPE developed an engineer's creed in 1954. From its website, the creed is as follows:

"As a Professional Engineer, I dedicate my professional knowledge and skill to the advancement and betterment of human welfare.

I pledge:

- *To give the utmost of performance;*
- *To participate in none but honest enterprise;*
- *To live and work according to the laws of man and the highest standards of professional conduct;*
- *To place service before profit, the honor and standing of the profession before personal advantage, and the public welfare above all other considerations.*

In humility and with need for Divine Guidance, I make this pledge."

The NSPE COE flows from this creed. As an organization dedicated to professional engineers, it covers all disciplines. Familiar highlights are:

- Provide utmost performance
- Participate in honest enterprises
- Live in accordance with the highest standards of professional conduct (this is nebulous at best, but can be used in lawsuits and also why there is *never* just one engineer on a project)
- Service before profit
- Honor and standing above personal advantage
- Health and welfare above all else

Further, professional engineers are required to:

- Notify the client when judgment is overruled
- Sign and seal only in areas of competency

- Report violations in the code of conduct
- Avoid conflicts of interest
- Refuse outside compensation
- Review plans only when requested to do so and advise the standing engineer of the review and why

In reviewing this history, it can be seen that licensure comes from codes, which came from guilds, which came from the collective expectations of the public. The engineer has a civic duty, as Kant notes, to protect the public. However, the public does not need to know exactly what an engineer does; the public needs only have the confidence that the engineer is competent to perform the work correctly. Licensure is that notice to the public that the engineer is indeed competent.

NSPE goes further with its COE. It notes that as members of the engineering profession, "engineers are expected to exhibit the highest standards of honesty and integrity." It provides six canons to direct the way (NSPE, 2020):

1. "Hold paramount the safety, health, and welfare of the public.
2. Perform services only in areas of their competence.
3. Issue public statements only in an objective and truthful manner.
4. Act for each employer or client as faithful agents or trustees.
5. Avoid deceptive acts.
6. Conduct themselves honorably, responsibly, ethically, and lawfully so as to enhance the honor, reputation, and usefulness of the profession."

This NSPE COE identifies certain issues that engineers should be cognizant of:

- "Engineers may express publicly technical opinions that are founded upon knowledge of the facts and competence in the subject matter.
- Engineers shall not accept compensation, financial or otherwise, from more than one party for services on the same

project, or for services pertaining to the same project, unless the circumstances are fully disclosed and agreed to by all interested parties.

- Engineers shall not solicit or accept financial or other valuable consideration, directly or indirectly, from outside agents in connection with the work for which they are responsible.
- Engineers shall not offer, give, solicit, or receive, either directly or indirectly, any contribution to influence the award of a contract by public authority, or which may be reasonably construed by the public as having the effect or intent of influencing the awarding of a contract.
- Engineers shall advise their clients or employers when they believe a project will not be successful.
- Engineers shall not accept outside employment to the detriment of their regular work or interest. Before accepting any outside engineering employment, they will notify their employers.
- Engineers shall not attempt to attract an engineer from another employer by false or misleading pretenses.
- Engineers shall not promote their own interest at the expense of the dignity and integrity of the profession.
- Engineers shall treat all persons with dignity, respect, fairness and without discrimination."

2.4 SUMMARY

So, what can be learned from this chapter? In the early years of civilization, the expectations of the tribe were important. Do your job, do it correctly, and you were allowed to remain in the tribe. As agriculture grew, the number of professions increased, but the same expectations remained. Ancient philosophers attempted to apply specific *rules* to behavior, but were only partially successful. They did, however, leave a legacy of concepts that were useful in defining those actions within public expectations. These expectations continued into the professional societies throughout the nineteenth century as the engineer's professional role became more defined. The result led to licensure, the subject of the following chapter.

REFERENCES

Anderson, R. Lanier. (2017). Friedrich Nietzsche. *Stanford Encyclopedia of Philosophy*. Metaphysics Research Lab. Stanford University.

Anscombe, G.E.M. (1958). Modern Moral Philosophy. *Philosophy, 33*(124): 1–19.

Aristotle. (1975). Nicomachean ethics (Sir David Ross, Trans.). Oxford University Press. London, UK.

ASCE. (2009). *Achieving the Vision for Civil Engineering 2025: A Roadmap for the Profession*. ASCE. Reston, VA.

Athanassoulis, N. (2000). A Response to Harman: Virtue Ethics and Character Traits. *Proceedings of the Aristotelian Society* (New Series). 100: 215–21.

Barbour, I. Philosophy and Human Values. In D. Douglas, G. Papadopoulos, and J. Boutelle. (2010). *Citizen Engineer: A Handbook for Socially Responsible Engineering*. Prentice Hall. Upper Saddle River, NJ.

Bentham, Jeremy. (1789). *Introduction to the Principles of Morals and Legislation*.

Bentham, Jeremy. (2001). *The Works of Jeremy Bentham: Published under the Superintendence of His Executor, John Bowring*. Volume 1. Adamant Media Corporation. p. 18.

Bowen, W.R. (2009). *Engineering Ethics: Outline of an Aspirational Approach*. Springer-Verlag. London, UK.

Brewer, Talbot. (2009). *The Retrieval of Ethics*. Oxford University Press. New York, NY.

Donaldson, Thomas and Thomas Dunfee. (1999). *Ties that Bind*. Harvard Business Review Press.

Fitzpatrick, K. and C. Gauthier. (2001). Toward a Professional Responsibility Theory of Public Relations Ethics. *Journal of Mass Media Ethics, 16*(2&3): 193–212.

Fledderman, C.B. (2004). *Engineering Ethics.* 2nd ed. Pearson Education. Upper Saddle River, NJ.

Friedman, M. (1982). *Capitalism and Freedom.* 2nd ed. The University of Chicago. Chicago, IL.

Hobbes, Thomas. (1658). *Elements of Philosophy.*

Hume, D. (1751). *An Enquiry Concerning the Principles of Morals.*

Hursthouse, R. and G. Pettigrove. (2016). *Virtue ethics.* In Edward N. Zalta, ed., *The Stanford Encyclopedia of Philosophy.* Metaphysics Research Lab. Stanford University. Winter 2016 ed.

Kant, Immanuel. (1785). First Section: Transition from the Common Rational Knowledge of Morals to the Philosophical. *Groundwork of the Metaphysics of Morals.*

Kant, Immanuel (1785a). Thomas Kingsmill Abbott (ed.). *Fundamental Principles of the Metaphysic of Morals* (10th ed.). Project Gutenberg. p. 23.

Kawall, Jason. (2009). In Defence of the Primacy of Virtues. *Journal of Ethics and Social Philosophy, 3*(2): 1–21.

Kraut, Richard. (1989). *Aristotle on the Human Good.* Princeton University Press. Princeton, NJ.

MacIntyre, Alasdair. (1985). *After Virtue.* 2nd ed. Duckworth. London, UK.

Mackie, J.L. (1977). *Ethics: Inventing Right and Wrong.* Penguin. London, UK:

Mantell, M.L. (1964). *Ethics and Professionalism in Engineering.* Collier-MacMillan Ltd. London, UK.

Mill, J.S. (1843). *A System of Logic.*

———. (1961/1863). Utilitarianism. In *Fraser's Magazine.*

Moore, G.E. (1903). *Principia Ethica.*

NSPE. (2020). Engineer's Creed. https://www.nspe.org/resources/ethics/code-ethics/engineers-creed. Accessed 12/27/20.

Perry, David. (1992). From S. Robinson, R. Dixon, C. Preece, and K. Moodley. (2007). *Engineering, Business and Professional Ethics.* Butterworth-Heinemann. Burlington, MA.

Rand, A. (1964). *The Virtue of Selfishness.* Signet.

Ross, W.D. (1927). The Basis of Objective Judgements in Ethics. *International Journal of Ethics*. p. 37.

Schwartz, J. (2016). Integrity: the Virtue of Compromise. *Palgrave Communications*. 2, 16085. https://doi.org/10.1057/palcomms.2016.85.

Smith, A. (1776/1952). *An Inquiry into the Nature and Causes of the Wealth of Nations*. p. 55. University of Chicago Press. Chicago, IL.

Stanford University. (2004). Ancient Ethical Theory. *Stanford Encyclopedia of Philosophy*. https://plato.stanford.edu/entries/ethics-ancient/#9.

Swanton, Christine. (2003). *Virtue Ethics: A Pluralistic View*. Oxford: Oxford University Press.

Watson, Gary. (1990). *On the Primacy of Character*. In Flanagan and Rorty. pp. 449–83. Reprinted in Statman, 1997.

Zagzebski, L. (1996). *Virtues of the Mind*. Cambridge University Press, New York, NY.

———. (2004). *Divine Motivation Theory*. Cambridge University Press. New York, NY.

PROBLEMS

1. A professional engineer (A) is asked to review a project where the client believes the cracks in the structural tank walls were caused by a design deficiency from engineer (B). Engineer (A) is asked to tell the client if engineer (B) is at fault. What is the proper course of action for engineer (A)?
2. Review any ASCE Journal. Pick an ethics issue from the past three years. Discuss the case and what the ethics conclusion was.
3. The principle of ethics goes back 10,000 years.
   a. True
   b. False

4. The Greeks first came up with the idea of ethics.
    a. True
    b. False
5. Ethics developed from the expectations of members of the tribe, village, and society.
    a. True
    b. False
6. Tribal ethics that would be respected include:
    a. Thievery
    b. Hunting skills
    c. Selfishness
    d. Disrespecting the elders
7. Apprenticeship requires:
    a. Training
    b. Education
    c. Experience
    d. All of these
8. The question *If everybody did ____________, would society function?* is a good starting spot to ask when confronted by an ethical dilemma.
    a. True
    b. False
9. Utilitarianism is an ancient philosophy that deals with greed and apportionment.
    a. True
    b. False
10. Utilitarianism assumes:
    a. A sense of duty
    b. Education and training
    c. Maximizing the happiness of people
    d. Being objective

11. The highest priority from an engineer's perspective according to NSPE should be the:
    a. Client
    b. Firm
    c. Public
    d. Law
12. The lowest priority from an engineer's perspective according to NSPE should be the:
    a. Client
    b. Firm
    c. Public
    d. Law
13. Engineers should not:
    a. Notify the client when their judgment is overruled
    b. Report violations in the code of conduct
    c. Accept outside compensation
    d. Put service before profit
14. The concept of protecting the profession suggests that engineers should never:
    a. Market their services
    b. Notify the client when their judgment is overruled
    c. Report violations in the code of conduct
    d. Lobby elected officials for work
15. Ethics for engineers do not come from:
    a. The concept of apprenticeships
    b. Codes
    c. Canons
    d. Due diligence
16. Virtue ethics are defined by values and behaviors that will allow a person to experience happiness and well-being.
    a. True
    b. False

17. Virtue ethics involve the principles of Aristotle.
    a. True
    b. False
18. The concept of *sustainability* has been added as a value of engineering ethics over the past 20 years.
    a. True
    b. False
19. Sustainability is part of utilitarianism.
    a. True
    b. False
20. Virtue ethics emphasizes the importance of and obligations to the public and clients.
    a. True
    b. False
21. Ethical norms do not change with time.
    a. True
    b. False

This book has free material available for download from the Web Added Value™ resource center at *www.jrosspub.com*

CHAPTER 3

STATES AND LICENSURE

The previous chapter described where the concept of ethics came from, why licensure developed, and the responsibilities that went along with both concepts. It was noted that the concept of licensure stems from a responsibility to the citizens and the expectation that engineers will act to protect public interests.

When you ask people what kinds of qualities they admire in others or in themselves, the answers commonly given are honesty, caring, fairness, courage, perseverance, diligence, trustworthiness, and integrity. The qualities on this list have something in common—they are distinctively ethical characteristics.

The push for engineering licensure came at the turn of the last century in the west, specifically because of a number of railroad bridge failures that negatively affected the public trust. Today in the United States, engineers and surveyors are licensed at the state and territory level. The U.S. model has generally only required the licensure of practicing engineers who are offering engineering services that impact the public welfare, safety, safeguarding of life, health, or property to be licensed.

LEARNING OBJECTIVES

- Understand the history of licensure in the United States
- Know the requirements for licensure
- Know the requirements to maintain a license
- Relate recent efforts to elevate the ethics discussion

- Understand what the consequences are regarding unethical behavior
- Identify ethical issues that might arise within a career
- Understand how to evaluate various ethical situations and options to confront them
- Know what role licensure plays in ethics
- Define and understand the impacts of misconduct to licensure
- Outline what actions involve ethical concerns and the disciplinary actions that boards can take

3.1 LICENSURE

Society will always be expected to undertake the construction of capital projects to erect buildings or install the supporting infrastructure needed for economic development. These projects include water mains, sewer lines, stormwater drainage, transportation, parks, and power lines as well as developing mechanical equipment to operate these systems. Proper design of transportation systems and control systems will also fall under these responsibilities. Society benefits from the constant improvement to technology in order to improve efficiency and productivity, with the new tools and gadgets made available for widespread use.

Developing infrastructure and technology is the purview of the engineer. For efficient operation, these newly constructed facilities must be developed in accordance with the latest technical and professional standards to protect the health, safety, and welfare of the customers being served now—and in the future—which requires engineers to keep pace with the latest technological advancements.

It is the engineer's responsibility to identify these standards and technologies, and to ensure that all designs meet or exceed any standards, codes, or other requirements that may apply to their development or use. Licensure is designed to indicate and certify that the specific persons doing the work understand the public health, safety, and welfare issues; are competent to design the project at hand; and understand the tools, equipment, materials, and other factors that are necessary to bring a project to a successful conclusion.

Licensure in the United States began in 1907 when Wyoming required certification for both engineers and surveyors in response to railroad trestle accidents in the state. By 1943, every state had created a licensing board. As all 50 states enacted licensure legislation, they began to see a need for a national council to help improve uniformity of laws and to promote mobility of licensure. The National Council of Examiners for Engineering and Surveying (NCEES) is a nonprofit organization that is dedicated to advancing professional licensure for engineers and surveyors that was created in 1920 for these very reasons. One of its missions is to support the state licensing boards with respect to competency and the mobility among the licensing jurisdictions by ensuring basic engineering competency via two exams.

NCEES provides the standardized Fundamentals of Engineering (FE) and Principles and Practice of Engineering (PPE) exams, both of which are required in order to obtain licensure. In addition, many states require that the degrees received by engineers are accredited by a specific review body to ensure that basic requirements are met. That review body is the Accreditation Board for Engineering and Technology (ABET), which will be discussed in the next chapter.

Today, obtaining that professional engineering license generally requires having an undergraduate degree in engineering as a prerequisite, passing the FE and PPE licensing exams, and having four years of progressively responsible experience that demonstrates competency in designing a project. References, generally from other professional engineers, are secured to verify any candidate's qualifications, and the licensing board reviews each individual applicant.

While every state had created a licensing board by 1943, it would be another 40 years before the requirements among all states were made relatively uniform. As a result, not all engineers are uniformly licensed, and it is important to make a distinction between a *graduate engineer* and a *professional engineer*.

A graduate engineer is a person who holds a degree in engineering from a four-year university program but is not licensed to practice or offer services to the public. A professional engineer (P.E.) is licensed to practice engineering in the jurisdiction. Note that having a P.E. license does not permit you to use "P.E." after your name in a state in which you are not licensed. Hence, all engineers must obtain licensure in the state or territory in which they are working in order to avoid a legal conflict.

Further, in some states, only those who are licensed are permitted to specifically identify themselves as engineers. For example, in Florida, only engineers who are licensed in the State of Florida are permitted to use the terms *civil engineer, mechanical engineer, chemical engineer*, or about 80 other professional descriptions (Florida Statute 471).

Table 3.1 highlights the basic inception dates and requirements of each of the 50 state licensing boards. Holding a P.E. license demonstrates to the public that the license holder has obtained the requisite education, experience, and knowledge necessary to make engineering judgments, and that the public can rely upon that person to protect their health, safety, and welfare. Most states renew licenses every two years and require continuing education to be obtained during that period. The most common requirement is 24 or 30 hours of contact time while in training. Today, that training can be obtained online or in person. Successful completion of online training involves successful completion of an exam on the online material presented.

Every engineering licensing board has its own laws within its jurisdiction. In general, the candidate must: (1) earn a degree from an ABET-accredited engineering program, (2) pass the FE exam, (3) gain four years of progressively responsible engineering work experience under the supervision of a professional engineer, and (4) pass the PPE exam in the appropriate discipline. If the candidate accomplishes these four requirements, most every state will permit a comity or reciprocal license to be obtained without retaking the PPE exam. However, where a state may not require an ABET accreditation, comity or reciprocity is not a given.

Table 3.1 Summary of licensing boards for professional engineers (data derived from NCEES (2019) and searches of licensing board websites). The 4th column, Cases over 5 years, represents the number of disciplinary cases each state has heard. The "X's" then represent the most common type of disciplinary case heard (CEU violation, No License violation, and Negligence violation), if known.

State	Registered Engineers	Year Incorporated	Cases over 5 years	CEU viol	No Lic Viol	Negl Viol	CEUs/2 yrs
Alabama	16,389	1935	83	X			30
Alaska	5,560	1939	6			X	0
Arizona	19,449	1921	285		X		0
Arkansas	9,209	1938	7			X	30
California	95,800	1920	70				0
Colorado	27,799	1919	120				0
Connecticut	11,045	1949	NR				24
Delaware	7,152	1941	NR				24
Florida	40,330	1917	119	X			18
Georgia	21,258	1937	13				30
Hawaii	7,445	1923	16				0
Idaho	7,906	1919	84				30
Illinois	24,070	1917	107				30
Indiana	13,498	1935	5				30
Iowa	9,485	1919	53			X	30

Continued

Table 3.1 *(continued)*

State	Registered Engineers	Year Incorporated	Cases over 5 years	CEU viol	No Lic Viol	Negl Viol	CEUs/2 yrs
Kansas	12,197	1931	11				30
Kentucky	13,886	1938	127				30
Louisiana	17,772	1908	283				30
Maine	16,346	1935	132				30
Maryland	20,555	1937	NR				16
Massachussetts	6,729	1934	NR				0
Michigan	20,531	1919	NR				30
Minnesota	14,622	1941	76		X		24
Mississippi	17,606	1928	62			X	30
Missouri	10,989	1941	220				30
Montana	6,262	1947	36				30
Nevada	11,800	1937	126				30
Nebraska	8,409	1935	41				30
New Hampshire	6,036	1945	5		X		30
New Jersey	18,623	1921	145				30
New Mexico	9,024	1935	6		X		30
New York	30,702	1920	NR				24

Continued

Table 3.1 *(continued)*

State	Registered Engineers	Year Incorporated	Cases over 5 years	CEU viol	No Lic Viol	Negl Viol	CEUs/2 yrs
North Carolina	27,923	1921	NR				30
North Dakota	5,484	1941	NR				30
Ohio	25,716	1933	579				30
Oklahoma	12,089	1935	NR				30
Oregon	14,303	1919	23		X		30
Pennsylvania	20,108	1919	25				24
Rhode Island	4,894	1938	7				0
South Carolina	18,146	1922	351		X		30
South Dakota	4,690	1925	11		X		30
Tennessee	13,251	1921	NR				24
Texas	58,299	1937	375			X	24
Utah	10,897	1929	NR				30
Vermont	4,175	1929	54				30
Virginia	29,351	1920	220	X			16
Washington	27,245	1937	35		X		0
West Virginia	8,933	1921	87		X		30
Wisconsin	14,280	1932	58		X		30
Wyoming	7,519	1907	21	X			30

Most states require completion of a specific number of hours of class time for each license renewal cycle. Continuing education curriculum is designed to allow engineers to keep up with new technology and changes to relevant building and design codes. The most common issues that arise when maintaining licensure are a failure to obtain the contact hours, negligence, and unlicensed people attempting to perform the required engineering work.

3.2 RULES GUIDING ENGINEERS

Protection of the public health, safety, and welfare is a public trust issue that can be equated with the expectations of doctors, firemen, and police officers since the motivations among these practitioners are similar. Holding a P.E. license demonstrates to the public that the professional has obtained the requisite education, experience, and knowledge that is necessary to make engineering judgments, and that the citizenry can rely upon that person to protect the health, safety, and welfare of the public at large. Having a P.E. license allows the professional to perform engineering consulting, own their own businesses, and bid for public funding.

Rules for each state may change periodically. Most states send out a newsletter or notices that contain proposed rule changes (often intended for public comment). Pay attention to these rules, as they have impact upon the profession. All engineers should have access to the rules for their state(s).

There are several issues that continually arise across state lines beyond comity/reciprocity. One issue is that if an engineer has licenses in multiple states, discipline issues that arise in one state are likely to occur in another. The ethics class requirements for one state may not be sufficient to be recognized in another. Online classes, certain topics, and certain providers may not be acceptable to all.

The engineer should always check with providers to ensure that they meet the requirements of a given state. Most professional society classes/courses and most university courses automatically qualify as long as they are related to engineering. Business and personal development classes rarely qualify in this manner.

If an engineer holds a license in one state, that is the only state in which he or she can practice. If the engineer moves to another state, he or she cannot use "P.E." after their name until such time as licensure has been granted in the new state. Many engineers are so used to putting "P.E." after their name that they do this as a habit without understanding the consequences.

3.3 FAILING TO DO YOUR JOB

From a legal perspective there are three basic areas of failure associated with engineering errors. These are:

- Negligence
- Incompetence
- Misconduct

Negligence is defined as the failure to exercise due care in the performance of the work *or* something that an ordinarily prudent person would foresee as a risk of harm to others if not corrected. Negligence can constitute grounds for disciplinary action by the Board of Professional Engineers, but not criminal prosecution in most states.

Incompetence is defined as a lack of *ability* to perform a function or a lack of *qualification* to perform a function or a lack of physical or mental ability to perform. Like negligence, incompetence can constitute grounds for disciplinary action by the Board of Professional Engineers, but not criminal prosecution.

While negligence and incompetence are problematic, it is misconduct that creates the most clearly defined issues among the proceedings at licensing boards. Misconduct includes (among others) any of the following:

- Expressing opinion without facts
- Issuing untrue statements in reports

- Signing and sealing work not done under direct supervision
- Offering or accepting a bribe for work or specifying certain products
- Acting when not qualified
- Affixing a seal when lacking competence to do so
- Conflicts of interest
- Allowing judgment to be overruled by nonprofessionals

The first two arise periodically in disciplinary hearings. The reason is that in many circumstances where engineers might speak, the public expects that they know more about the issue than lay people. For example, when speaking about the location of a landfill, garbage transfer station, airport, or power plant, the decision makers may give more weight to what an engineer says. Therefore, uninformed opinions may hurt the profession and when these opinions are incorrect or exposed as having a particular agenda, discipline will follow.

Likewise, errors or incorrect statements in reports or findings that are submitted to clients and regulatory agencies or in a review of incidents can result in major penalties. Protection of the profession is the critical issue.

Most licensing boards list a series of transgressions, as well as penalties for each that range from small fines to license revocation. Table 3.2 outlines the options for Florida. There may be extenuating circumstances that lessen the severity of any issue—and the opposite is also true.

Having discussed the causes and consequences of unethical behavior, let us look at some case studies. Most of these are relatively straightforward, and all represent actual cases in a variety of states across the United States.

Table 3.2 Examples of violations and penalties in Florida for engineering rules violations

Violation	Penalty Range	
	First Violation	**Second and Subsequent Violations**
(a). Violating any provision of Section 455.227(1), 471.025, or 471.031, F.S., or any other provision of Chapter 471, F.S., or rule of the Board or Department. (Sections 471.033(1)(a) and 455.227(1)(b), (q), F.S.)	Reprimand and $1,000 fine; to one (1) year suspension, two (2) years probation and $5,000 fine.	One (1) year suspension, two (2) years probation, and $5,000 fine; to revocation.
1. Failure to sign, seal, or date documents. (Section 471.025(1), F.S.)	Reprimand to one (1) year probation.	Reprimand and one (1) year probation to revocation.
2. Sealing any document after license has expired or been revoked or suspended, or failure to surrender seal if the license has been revoked or suspended. (Section 471.025(2), F.S.)	Suspended license: revocation and $1,000 fine. Revoked license: referral to State Attorney's Office.	Suspended license: revocation and $5,000 fine. Revoked license: referral to State Attorney's Office.
3. Signing or sealing any document that depicts work the licensee is not licensed to perform or which is beyond his or her profession or specialty therein or practicing or offering to practice beyond the scope permitted by law or accepting and performing responsibilities the licensee is not competent to perform. (Sections 471.025(3), 455.227(1)(o), F.S., paragraphs 61G15-19.001(6)(c), (d), F.A.C.)	Reprimand, one (1) year probation and $1,000 fine; to $5,000 fine, one (1) year suspension, and two (2) years probation.	Reprimand, $5,000 fine, one (1) year suspension, and two (2) years probation; to revocation.
4. Firm practicing without proper qualification. (Section 471.023, F.S., and subsection 61G1519.001(3), F.A.C.)	$1,000 fine; to $5,000 fine.	$5,000 fine.

Continued

Table 3.2 *(continued)*

Violation	Penalty Range	
	First Violation	**Second and Subsequent Violations**
5. Practicing engineering without a license or using a name or title tending to indicate that such person holds an active license as an engineer. (Sections 471.031(1)(a), (b), F.S.)	$1,000 fine; to $5,000 fine.	$5,000 fine and referral to State Attorney's Office.
6. Presenting as his or her own the license of another. (Section 471.031(1)(c), F.S.)	$1,000 fine; to $5,000 fine.	$5,000 fine and referral to State Attorney's Office.
7. Giving false or forged evidence to the Board or concealing information relative to violations of this chapter. (Sections 471.031(1)(d), (g), F.S.)	$1,000 fine; to $5,000 fine and suspension.	Reprimand and $5,000 fine; to revocation.
8. Employing unlicensed persons to practice engineering or aiding, assisting, procuring, employing unlicensed practice or practice contrary to Chapter 455 or 471, F.S. (Sections 471.031(1)(f), and 455.227(1)(j), FS)	$1,000 fine and reprimand; to $5,000 and suspension.	Reprimand and $5,000 fine; to revocation.
9. Having been found liable for knowingly filing a false complaint against another licensee. (Section 455.227(1)(g), F.S.)	$1,000 fine and reprimand; to $5,000 per count and suspension.	Reprimand and $5,000 fine; to revocation.
10. Failing to report a person in violation of Chapters 455 and 471, F.S., or the rules of the Board or the Department. (Section 455.227(1)(i), F.S.)	Reprimand; to $5,000 and suspension for one (1) year.	Reprimand and $5,000 fine; to revocation.

Continued

Table 3.2 *(continued)*

Violation	Penalty Range	
	First Violation	**Second and Subsequent Violations**
11. Failing to perform any statutory or legal obligation. (Section 455.227(1)(k), F.S.)	Reprimand; to one (1) year suspension and a $1,000 fine.	Reprimand and $5,000 fine; to revocation.
12. Exercising influence on a client for financial gain. (Section 455.227(1)(n), F.S.)	Reprimand; to one (1) year suspension and $5,000 fine.	Reprimand and $5,000 fine; to revocation.
13. Improper delegation of professional responsibilities. (Section 455.227(1)(p), F.S.)	$1,000 fine and probation for one (1) year; to suspension.	Reprimand and $5,000 fine; to revocation.
14. Improperly interfering with a disciplinary investigation or inspection or proceeding. (Section 455.227(1)(r), F.S.)	$1,000 fine and probation for one (1) year; to suspension.	Reprimand and $5,000 fine; to revocation.
(b). Attempting to procure a license by bribery, fraudulent misrepresentation, or error of the Board or Department. (Sections 471.033(1)(b) and 455.227(1)(h), F.S.)	One (1) year suspension and $1,000 fine; to revocation if licensed; if not licensed, denial of license and referral to State Attorney's Office.	Revocation and $5,000 fine if licensed; if not licensed, denial of license and referral to State Attorney's Office.
(c). Having a license to practice engineering acted against or denied by another jurisdiction. (Sections 471.033(1)(c) and 455.227(1)(f), F.S.)	Same penalty as imposed in other jurisdiction or as close as possible to penalties set forth in Florida Statutes.	Same penalty as imposed in other jurisdiction or as close as possible to penalties set forth in Florida Statutes.
(d). 1. Being convicted or found guilty of, or entering a plea of nolo contendere to a crime which relates to the practice or ability to practice. (Sections 471.033(1)(d) and 455.227(1)(c), F.S.)	Depending on the severity of the crime, from reprimand; $1,000 fine, and one (1) year probation; to revocation.	Depending on the severity of the crime, from one (1) year suspension with two (2) years probation; to revocation.

Continued

Table 3.2 *(continued)*

Violation	Penalty Range	
	First Violation	**Second and Subsequent Violations**
(d). 2. Conviction of crime related to building code inspection or plans examination. (Paragraph 61G15-19.001(7)(a), F.A.C.)	Reprimand, $1,000 fine, and one (1) year probation.	One (1) year suspension with two (2) years probation; to revocation.
(e). Knowingly making or filing a false report or record, failing to file a report or record required by law, impeding or obstructing such filing. (Sections 471.033(1)(e), 455.227(1)(l), F.S., and paragraph 61G15-19.001(7)(c), F.A.C.)	Reprimand and $1,000 fine; to one (1) year suspension and two (2) years probation.	One (1) year suspension, two (2) years probation, and $2,000 fine; to revocation and $5,000 fine.
(f). Fraudulent, false, deceptive, or misleading advertising. (Sections 471.033(1)(f), F.S., and subsection 61G15-19.001(2), F.A.C.)	Reprimand to one (1) year probation and $5,000 fine.	One (1) year probation and $5,000 fine; to revocation.
(g). Fraud, deceit, negligence, incompetence, or misconduct. (Sections 471.033(1)(g) and 455.227(1)(a), (m), F.S.)		
1. Fraud or deceit	Reprimand, two (2) years probation and $1,000 fine; to one (1) year suspension and $5,000 fine.	One (1) year suspension and $5,000 fine; to revocation.

Continued

Table 3.2 *(continued)*

Violation	Penalty Range	
	First Violation	**Second and Subsequent Violations**
2a. Negligence. (Subsection 61G15-19.001(4), F.A.C.)	Reprimand, two (2) years probation and $1,000 fine; to $5,000 fine, five (5) year suspension, and ten (10) years probation.	Two (2) years probation and $1,000 fine; to $5,000 fine and revocation.
2b. Negligence in procedural requirements. (Subsections 61G15-30.003(2), (3) and (5), F.A.C, Rules 61G15-30.005 and 61G15-30.006, F.A.C.)	Reprimand; to two (2) years probation and $1,000 fine.	Two (2) years probation and $1,000 fine; to $5,000 fine and revocation.
2c. As a special inspector. (Subsection 61G15-19.001(5), F.A.C.)	Reprimand, two (2) years probation, and $1,000 fine; to $5,000 fine.	Two (2) years probation and $1,000 fine; to $5,000 fine and revocation.
3. Incompetence. (Subsection 61G15-19.001(5), F.A.C.)	Two (2) years probation; to suspension until ability to practice proved, followed by two (2) years probation.	Suspension until ability to practice proved, followed by two (2) years probation; to revocation.
4. Misconduct. (Subsection 61G15-19.001(6), F.A.C.)	Reprimand and $1,000 fine; to one (1) year suspension.	One (1) year suspension; to revocation and $5,000 fine.
a. Expressing an opinion publicly on an engineering subject without being informed as to the facts and being competent to form a sound opinion. (Paragraph 61G15-19.001(6)(a), F.A.C.)	Reprimand and $1,000 fine; to one (1) year suspension.	One (1) year suspension; to revocation and $5,000 fine.

Continued

Table 3.2 *(continued)*

Violation	Penalty Range	
	First Violation	**Second and Subsequent Violations**
b. Being untruthful, deceptive or misleading in any professional report, statement, or testimony or omitting relevant and pertinent information from such report, statement, or testimony when the result or such omission would or reasonably could lead to a fallacious conclusion. (Paragraph 61G15-19.001(6)(b), F.A.C.)	Reprimand and $1,000 fine; to one (1) year suspension.	One (1) year suspension; to revocation and $5,000 fine.
c. Offering directly or indirectly any bribe or commission or tendering any gift to obtain selection or preferment for engineering employment other than the payment of the usual commission for securing salaried positions through licensed employment agencies. (Paragraph 61G15-19.001(6)(e), F.A.C.)	Reprimand, $5,000 fine per count, and suspension for five (5) years; to revocation.	Five (5) years suspension; to revocation.
d. Soliciting or accepting gratuities without client knowledge. (Paragraphs 61G15-19.001(6)(g), (h), F.A.C.)	Reprimand, one (1) year probation, and $1,000 fine; to one (1) year suspension, two (2) years probation, and $5,000 fine.	One (1) year suspension, two (2) years probation, and $5,000 fine; to revocation.
e. Failure to preserve client's confidence. (Paragraph 61G15-19.001(6)(r), F.A.C.)	Reprimand, one (1) year probation, and $1,000 fine; to one (1) year suspension, two (2) years probation (if pecuniary benefit accrues to engineer).	One (1) year suspension, two (2) years probation, and $5,000 fine; to revocation.

Continued

Table 3.2 *(continued)*

Violation	Penalty Range	
	First Violation	**Second and Subsequent Violations**
f. Professional judgment overruled by unqualified person. (Paragraph 61G15-19.001(6)(l), F.A.C.)	Reprimand, one (1) year probation, and $1,000 fine; to one (1) year suspension, two (2) years probation, and $5,000 fine.	One (1) year suspension, two (2) years probation, and $5,000 fine; to revocation.
g. Use of name/firm in fraudulent venture. (Paragraph 61G15-19.001(6)(k), F.A.C.)	Reprimand, one (1) year probation, and $1,000 fine; to $5,000 fine, one (1) year suspension, and two (2) years probation.	One (1) year suspension, two (2) years probation, and $5,000 fine; to revocation.
h. Undisclosed conflict of interest. (Paragraphs 61G15-19.001(6)(f), (p), F.A.C.)	Reprimand, $1,000 fine, and two (2) years probation; to revocation and $5,000 fine.	One (1) year suspension, two (2) years probation, and $5,000 fine; to revocation.
i. Renewing or reactivating a license without completion of continuing education hours. (Paragraph 61G15-19.001(6)(s), F.A.C.)	Reprimand and $1,000 fine; to suspension until licensee demonstrates compliance.	One (1) year suspension and $1,000 fine; to revocation.
(h). Violating any provision of Chapter 455, F.S. (Sections 471.033(1)(h) and 455.227(1)(q), F.S.)	Depending on the severity of the violation, reprimand and $1,000 fine per count; to $5,000 fine and revocation.	Depending on the severity of the violation, one (1) year suspension, two (2) years probation, and $5,000 fine; to revocation.

Continued

Table 3.2 *(continued)*

Violation	Penalty Range	
	First Violation	**Second and Subsequent Violations**
(i). Practicing on a revoked, suspended, inactive or delinquent license, or through a business organization not properly qualified. (Sections 471.033(1)(i) and 471.031(1)(e), F.S.)		
1. Delinquent license.	Fine based on length of time in practice while inactive; $100/month or $1,000 maximum, renewal of license or cease practice.	
2. Inactive license.	Fine based on length of time in practice while inactive; $100/month or $1,000 maximum, renewal of license or cease practice.	
3. Suspended license.	Revocation and $1,000 fine.	
4. Revoked license.	Referral to State Attorney's Office.	Referral to State Attorney's Office.
5. Business Organization not properly qualified.	Reprimand, $500 fine; to $5,000 fine and one (1) year suspension.	One (1) year suspension and $5,000 fine; to revocation.

Continued

Table 3.2 *(continued)*

Violation	Penalty Range	
	First Violation	**Second and Subsequent Violations**
(j). Affixing or permitting to be affixed his or her seal, name, or digital signature to any documents that were not prepared by him or her or under his or her responsible supervision, direction or control. (Section 471.033(1)(j), F.S., and paragraphs 61G15-19.001(6)(j), (q), F.A.C.)	Reprimand, one (1) year probation, and $1,000 fine; to $5,000 fine, one (1) year suspension, and two (2) years probation.	One (1) year suspension, two (2) years probation, and $5,000 fine; to revocation.
(k). Violating any order of the board or department. (Sections 471.033(1)(k), 455.227(1)(q), F.S., and paragraph 61G1519.001(6)(o), F.A.C.)	Depending on the severity of the violation, from suspension until compliant with the order of the Board and $1,000 fine; to revocation and $5,000 fine.	Depending on the severity of the violation, suspension until compliant with the order of the Board and $1,000 fine; to revocation and $5,000 fine.
(l). Aiding, assisting, procuring, employing unlicensed practice or practice contrary to Chapter 455 or 471, F.S. (Section 455.227(1)(j), F.S.)	$1,000 fine and probation for one (1) year; to $5,000 fine and suspension.	Reprimand and $5,000 fine; to revocation.
(m). Failing to report in writing a conviction or plea of nolo contendere, a crime in any jurisdiction. (Section 455.227(1)(t), F.S.)	Reprimand; to $5,000 fine.	Six (6) month suspension; to $5,000 fine and revocation.

3.4 CASE STUDIES

The following sections describe some of the more common disciplinary issues that engineers have found themselves subjected to. A brief discussion of the rules and why they are in place is noted. Violation of a rule is *misconduct*. Practicing outside your area is also considered to be defined as misconduct. All examples of misconduct can be described as self-inflicted issues. Figure 3.1 outlines the process that

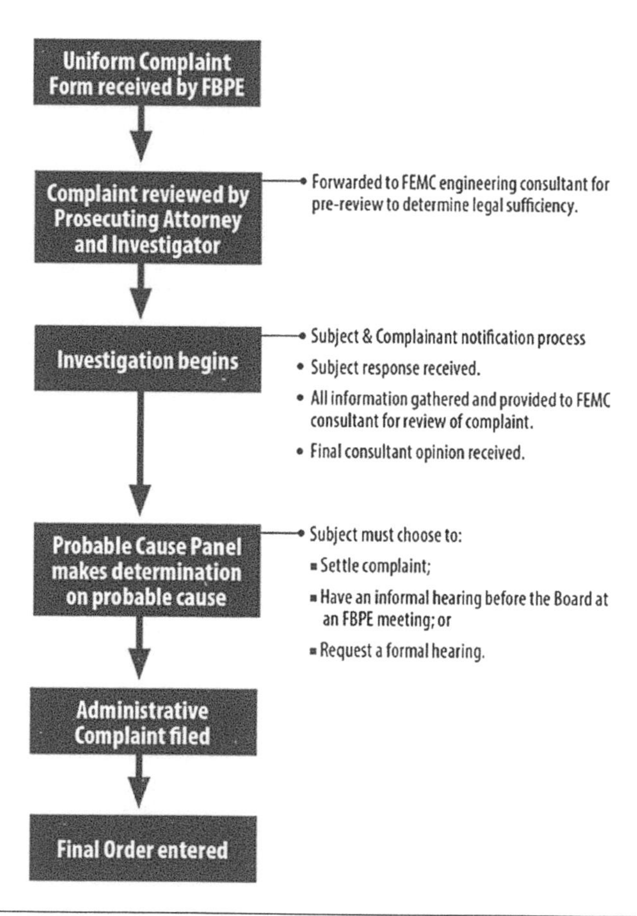

Figure 3.1 Process for addressing ethical or legal violations by engineers (*Source*: Florida Board of Professional Engineers at fbpe.org)

is generally used to review complaints. Note that a range of penalties, such as those shown in Table 3.2, can be applied.

3.4.1 Continuing Education

Continuing education is an important part of the engineering field. Technology changes, materials change, software changes, and other things change, thus the need for continuing education is vital. As outlined in this chapter, most states require some form of continuing education. Few states require that engineers submit proof to demonstrate compliance, but most states have text in their rules such as the following (fbpe.org):

- *"(1) In order to demonstrate compliance, licensees must attest to completion of the continuing education requirements upon licensure renewal. The Board will audit at random a number of licensees as is necessary to assure that the continuing education requirements are met.*
- *(2) The licensee shall retain such receipts, vouchers, certificates, or other papers as may be necessary to document completion of the continuing education pursuant to an audit for four years from the date of completion of the continuing education activity."*

Some states also require that continuing education providers be approved by the state board, or some agency such as a national society. The more significant issue is what happens if someone does not comply. After a recent renewal, one state found a significant percentage of engineers did not comply even though they signed a sworn statement that they did on their renewal form. This violates the National Society of Professional Engineers (NSPE) Canon 5 to *avoid deceptive acts* regarding any laws or rules in the applicable states. Failure to meet continuing education requirements is the most common violation in many of the states that require it. Two recent examples follow.

EXAMPLE 3.1

Florida (fbpe.org)

"J. C."

Licensee was charged with violating Section 471.033(1)(a), Florida Statutes and Rule 61G15-19.001(6)(s), Florida Administrative Code; renewing his Professional Engineer license without having completed all required continuing education. Licensee renewed his Professional Engineer license without having completed the Laws & Rules Course, the Ethics Course and 7.5 hours of General Engineering on or before February 28, 2017. Licensee did complete some of the outstanding Continuing Education courses. To date, licensee lacks one hour of Florida Laws & Rules, one hour of Engineering Ethics, and 5.5 hours of General Engineering.

RULING: The case was presented to the full Board upon a motion for determination that respondent forfeited his right to an administrative hearing. The Board imposed a suspension until licensee appears before the Board, an administrative fine of $1,000, and costs of $124.80. Final order was issued on July 16, 2019.

VIOLATION: Section 471.033(1)(a), Florida Statutes, and Rule 61G15-19.001(6)(s), Florida Administrative Code

EXAMPLE 3.2

North Carolina

"J. F."

VIOLATION: Failed to comply with annual CPC requirements [.1703, .1712] and failed to maintain adequate CPC records [.1706].

BOARD ACTION: Reprimand, ethics course and audit CPC the next three years renewed as current.

3.4.2 Licensure

One of the most common violations of the public confidence involves people who attempt to provide services for which they are not qualified. Realizing that engineers who are working for commercial manufacturing enterprises may not need a license to work (since the corporations are taking product liability responsibilities for faulty designs), there is a limitation on the number of engineers who are offering certain services that require a license. Licensure is a privilege, as noted earlier. A typical statute covering engineering licensure says: the privilege permits people who are covered (*by licensure*) to:

- Advertise that they provide engineering services
- Use letterhead and cards using engineering designations

In other words, if one were to use the designation *P.E.*, the public might infer that one has met the requirements for licensure in that state and that they are competent to perform the work. The ethical issue is derived from what the public perception is. As noted, because the public really does not understand what engineers do, there is a need for those in the profession to: "Conduct themselves honorably, responsibly, ethically, and lawfully, so as to enhance the honor, reputation, and usefulness of the profession manner" (per the NSPE code of ethics). Implicit in NSPE Canon 6 is that the profession and those within it should be protected.

NSPE also advises members to avoid deceptive acts. Hence, state engineering boards are likely to deal harshly with individuals who deceptively pass themselves off as engineers, thereby potentially endangering the public health, safety, and welfare since they are not qualified to perform such services.

Assume someone has an engineering degree and has worked for many years in construction. He or she applies for and gets a job with an agency where the job requirement is having—or at least having the ability to obtain—an engineering license, which the candidate claims to have. However, the candidate does not have those qualifications, but it takes a while for anyone to check whether this is true.

The result of this behavior is shown in Examples 3.3 and 3.4. Pretending to be an engineer violates Canon 5—avoid acts of deception—and the statutes in both states, which is why they were referred to the Attorney General for action. It also endangers the public health, safety, and welfare (NSPE Canon 1) because, in both cases, the candidates lacked the appropriate credentials.

EXAMPLE 3.3

North Carolina

"G. A."

VIOLATION: Practiced or offered to practice engineering without a license as required by G. S. 89C-23; presented or attempted to use the certificate of licensure or seal of another [G. S. 89C-23]; and falsely claimed to be licensed under G. S. 89C [G. S. 89C-23].

BOARD ACTION: Refer to State Bureau of Investigation.

EXAMPLE 3.4

Florida

"E. P."

VIOLATION: "E. P." was charged with one count of practicing engineering without a license, and one count of presenting another's license as his own. "E. P." failed to timely respond to the charges. Presentation to the full Board.

BOARD ACTION: Fined $10,000 plus costs of $380; referred to Florida Attorney General for further action.

Example 3.5 is a company that was passing itself off as an engineering company (which in most states requires that they provide a supervising engineer registered with the state). The individual in Example 3.5 was not that person. In this case the person signed and sealed deficient drawings under the name of an entity that was not an engineering firm, and for which he was not employed. Trouble occured when issues arose within the design.

EXAMPLE 3.5

Florida

"T. M."

The licensee was charged with one count for being misleading in the name of an unlicensed entity, the licensee is not an employee of the company that appeared on the licensee's letterhead and calculation sheets; one count of negligence relating to structural deficiencies in wind loading calculations, one count on aiding, assisting, procuring, employing, or advising an unlicensed person or entity to practice engineering.

The licensee entered into one stipulation with FEMC which imposed a $3,000,000 fine; costs of $1,673.03; a reprimand, a limitation from practicing structural engineering until such time as the licensee takes and passes the NCEES Structural II Examination, a two-year period of probation with project review; completion of a Board-approved course on professionalism and ethics; and completion of the Board's Study Guide.

What these three examples reveal is the lack of tolerance on the part of licensing boards for those who attempt to act as engineers but are not actually licensed to do so. As noted in Example 3.5, the penalties can be severe.

3.4.3 Failure to Seal Correctly

Yes, this is a *thing*. Sealing documents implies that they are final and original. The intent is to prevent someone from changing the work in the future by leaving the seal intact, making the prior engineer responsible. This is inappropriate. Most states have specific requirements on how signing and sealing occurs. These requirements extend to the seal, a signature, and the date. Some states have gone even further to require that applicable codes be identified and other notations. All this is done to make it easier for the people who are reviewing the plans to ensure that the correct guidelines are being used. A typical rule says:

> *"61G15-23.003 Procedures for Physically Signing and Sealing Plans, Specifications, Reports or Other Documents. Engineering plans, specifications, reports or other documents which must be" . . . " physically signed, dated, and sealed as provided herein by the professional engineer in responsible charge.*
>
> *(1) The licensee shall sign by hand an original of the licensee's signature on each page required to be sealed. A scanned, facsimile, digitally created, or copied image of the licensee's signature shall not be used.*
>
> *(2) The licensee must then use a wet seal, a digitally created seal, or an embossing seal placed partially overlapping the licensee's signature on each page required to be sealed. The placement of the seal shall not render the signature illegible."*

In Alabama:

> *"Ala. Admin. Code r. 330-X-11-.03—Seal On Documents. (1) The seal, signature, and date of signature on a document signify that the document was prepared by the licensee or under his or her responsible charge, or that the licensee has reviewed the document in sufficient depth to fully coordinate and assume responsibility*

for documents prepared by another licensed professional engineer or licensed professional land surveyor."

In addition, every sheet within the plans and prints is to be signed, dated and sealed by the professional engineer in responsible charge. To wit the typical requirements are:

"1. A title block shall be used on each sheet of plans or prints and shall contain the printed name, address, and license number of the engineer who has signed, dated, and sealed the plans or prints.

2. If the engineer signing, dating, and sealing engineering plans or prints is practicing through a duly authorized qualified engineering business organization; the title block shall contain the printed name and address of the qualified engineering business organization." (FL 61G15)

Reports and specifications should likewise be signed, sealed, and dated. A 2020 update in Florida requires:

"61G15-23.004 Procedures for Digitally Signing and Sealing Electronically Transmitted Plans, Specifications, Reports or Other Documents. . . .

"Engineering Specifications and Calculations. An index sheet shall be used and shall be signed, dated, and sealed by each professional engineer who is in responsible charge of any portion of the engineering specifications or calculations.

1. The index sheet must be signed, dated and sealed by those professional engineers in responsible charge of the production and preparation of each section of the engineering specifications or calculations, with sufficient information on the index sheet so that the user will be aware of each portion of the specifications or calculations for which each professional engineer is responsible.

2. The index sheet shall include at a minimum:

"a. The printed name, address, and license number of each engineer in responsible charge of the production of any portion of the calculations or specifications.

b. If the engineer signing, dating, and sealing calculations or specifications is practicing through a duly qualified engineering business organization; the printed name and address of the qualified engineering business organization.

c. Identification of the project, by address or by lot number, block number, section or subdivision, and city or county.

d. Identification of the applicable building code and chapter(s) and Florida Fire Prevention Code, when applicable, that the design is intended to meet.

e. Identification of any computer program used for engineering the specifications or calculations."

The computer programming issue arises from the fact that if the individual using the program does not understand how the program operates or calculates results, using the results in a design may introduce errors that cause flaws in the design.

So, what happens when someone does not sign and seal correctly or fails to provide the requisite information? A typical example would be where an individual is a civil engineer by training and does mostly subdivision work. Assume the engineer files several documents to be recorded in the Public Record. If the documents are not signed, sealed, and dated per the state rules, the Clerk of Courts (or some agency like that) would notify the Board that the documents had not been properly signed, sealed, and dated. What could happen? A typical penalty would be a fine of $1,000, payable within 30 days, with the suspension of licensure until the fine is paid if beyond 30 days. The state's licensing boards have numerous examples like this every year. It is one of the most common errors that is made.

3.4.4 Plan Stamping and Letting Someone Use Your Seal

Ultimately all rules developed by states require that engineers seal only those documents over which they exert responsibility. This is in keeping with NSPE Canon 2 which states: "Perform services only in areas of their competence." Plan stamping is not that. It also violates NSPE Canon 5: "Avoid deceptive acts." The deception is the engineer who is signing and sealing the plan has not actually prepared the document, but is acting as the *Engineer of Record* in responsible charge of the project. That can happen when stamping plans that are done by others for a fee (or even for no fee). The following are examples of rules states have adopted regarding signing and sealing:

> *"A professional engineer may only sign, date, and seal engineering plans, prints, specifications, reports, or other documents if that professional engineer was in responsible charge, as that term is defined in subsection 61G15-18.011(1), F.A.C., of the preparation and production of the engineering document, and the professional engineer has the expertise in the engineering discipline used in producing the engineering document(s) in question." (Florida 61G15)*

Michigan notes:

> *MCLA 339.2008(1) requires that all documents requiring governmental agency approval or record must be sealed by the person in responsible charge: "(1) A plan, plat, drawing, map, and the title sheet of specifications, an addendum, bulletin, or report or, if a bound copy is submitted, the index sheets of a plan, specification, or report, if prepared by a licensee and required to be submitted to a governmental agency for approval or record, shall carry the embossed, printed, or electronic seal of the person in responsible charge."*

> *MCLA 339.2008(3) provides: "(3) a licensee shall not seal a plan, drawing, map, plat, report, specification, or other document that is not prepared by the licensee or under the supervision of the licensee as the person in responsible charge."*

Colorado adds:

> *"CO 3.2.2—Seal and Sign Only Documents under Responsible Charge or Control. Licensees shall only affix their signatures and seals to plans or documents prepared under their responsible charge or control."*

Mississippi Rule 14.3 notes that:

> *"Each sheet of plans, drawings, documents, specifications, and reports for engineering practice and of maps, plats, and charts shall be signed, sealed, and dated by the licensee preparing them, prepared under his direct supervisory control, or reviewed by him in sufficient depth to fully coordinate and assume responsibility for documents that were prepared by another professional engineer."*

All of these rules use the requirement of responsible charge, which is defined as: *degree of control an engineer is required to maintain over engineering decisions made personally or by others over which the engineer exercises supervisory direction and control authority*. Michigan law defines the person in responsible charge of a project as follows:

> *MCLA 339.2001(d): "Person in responsible charge" means a person licensed under this article who determines technical questions of design and policy; advises the client; supervises and is in responsible charge of the work of subordinates; is the person whose professional skill and judgment are embodied in the plans, designs, plats, surveys, and advice involved in the services; and who supervises the review of material and completed phases of construction."*

Colorado has a much more extensive rule:

> *"Responsible Charge of Engineering. The Board shall interpret "responsible charge" of engineering, as defined in Section 12-25-102(14), C.R.S., as follows: "Responsible charge" of engineering shall mean that degree of control an engineer is required to maintain over engineering decisions made personally or by others over which the engineer exercises supervisory direction and control authority.*
>
> *(a) The degree of control necessary for an engineer to be in responsible charge shall be such that the engineer:*
>
> > *(i) Personally makes engineering decisions, or personally reviews and approves proposed decisions prior to their implementation, including consideration of alternatives whenever engineering decisions that could affect the life, health, property, and welfare of the public are made. In making said engineering decisions, the engineer shall be physically present or, through the use of communication devices, be available in a reasonable period of time as appropriate.*
> >
> > *(ii) Judges the validity and applicability of recommendations prior to their incorporation into the work, including the qualifications of those making the recommendations.*
>
> *(b) Engineering decisions that are made by, and are the responsibility of, the professional engineer in responsible charge are those decisions concerning permanent or temporary work that could create a danger to the life, health, property, and welfare of the public, such as, but not limited to, the following:*
>
> > *(i) The selection of engineering alternatives to be investigated and comparison of alternatives for engineering works.*
> >
> > *(ii) The selection or development of design standards or methods, and materials to be used.*

(iii) The selection or development of techniques or methods of testing to be used in evaluating materials or completed works, either new or existing.

(c) As a test to evaluate whether an engineer is in responsible charge the following must be considered. An engineer who signs and seals engineering documents in responsible charge must be capable of answering questions as to the engineering decisions made during the engineer's work on the project in sufficient detail as to leave little doubt as to the engineer's proficiency for the work performed. It is not necessary to defend decisions as in an adversary situation, but only to demonstrate that the engineer in responsible charge made them and possessed sufficient knowledge of the project to make them. Examples of questions to be answered by the engineer could relate to criteria for design, methods of analysis, selection of materials and systems, economics of alternate solutions, and environmental considerations. The individual should be able to clearly define the degree of control and how it was exercised and be able to demonstrate that the engineer was answerable within said degree of control necessary for the engineering work done.

(d) The term "responsible charge" does not refer to financial liability. (e) A professional engineer who adopts, signs, and seals work previously engineered shall perform sufficient review and calculation to ensure that all standards of practice required of licensees are met, including satisfying the relevant criteria stated in paragraphs (b) and (c) above, and shall take professional and legal responsibility for documents signed and sealed under his/her responsible charge."

Permitting someone to use your seal fails this test as well. Signing and sealing outside your area also fails these requirements as noted in Examples 3.6 and 3.7.

EXAMPLE 3.6

North Carolina

"C. R."

VIOLATION: Affixed seal to work not done under direct supervisory control or responsible charge [.0701(c)(3)] by certifying a structural assessment report without visual observation by the licensee or someone under direct supervisory control; and failed to obtain agreement in writing for payment from more than one party for services on the same project, or pertaining to the same project [.0701(e)(2)].

BOARD ACTION: Reprimand.

EXAMPLE 3.7

Florida

"E. C."

VIOLATION: The licensee was charged with one count of negligence for failure to safeguard his seal which was used by someone other than the licensee to seal several sets of plans for a residential project.

ACTION: The licensee entered into a stipulation with the FEMC, which imposed a $1,000 fine, costs of $209.63, a reprimand, one year probation, completion of the board-approved course on Professionalism and Ethics, and completion of the Board's Study Guide.

The same can be said for signing and sealing plans that are not final. If the plans are not final, signing and sealing them misrepresents their status to the public. Colorado notes that:

> *"6.1.4—Sealing Documents That Are Not Final. When a licensee seals surveying documents that are not final, the status of the*

surveying documents must be identified as preliminary. Further qualifying descriptors may be added, e.g., "for review."

State boards will discipline engineers who fail in this regard. Florida has a similar rule that was borne out of an incident where a permitting agency required signed and sealed plans submitted for permit evaluation but were not final (a fairly common occurrence).

EXAMPLE 3.8

Florida

"Mary"

Mary is a P.E. She signed, dated, and sealed some preliminary plans without indicating they were preliminary. What could happen? The seal means the plans meet all codes and rules. It also implies finality. Permit copies are not final. Hence, they fined her.

Actually, the case in Example 3.8 created a lot of discussion and altered the rules in Florida (see Figure 3.2).

Signed and Sealed Drawings that are not Final Drawings

The Florida Engineers Management Corporation continues to receive complaints against Engineers from Building Officials that signed and sealed plans submitted for permit (public record) are incomplete, incorrect, have required information missing, etc. After investigation by the FEMC staff, frequently the Engineer's defense is that "The plans were preliminary", "...for review", "...for financing", etc. The result is considerable time and expense expended by FEMC, the FBPE Probable Cause Panel, the FBPE Board, and the defendant's time and possible attorney fees.

Florida Statute 471.025 indicates that "all final drawings ...being filed for public record and all final bid documents ...shall be ...signed, dated, and stamped ...." (emphasis added). FBPE Rule 61-G15-23.002(4) expands the statute by indicating that preliminary plans not intended for permit, construction or bidding purposes should not be signed, sealed, and dated. However, if a permitting agency requires plans submitted for review to be signed and sealed, the engineer should clearly note limitations by placing on the plans such terms as "Preliminary", "For Review Only", Not for Construction", or any suitable statement indicating that the submittal is not for permit, construction or bidding.

Documents filed for permit (public record) shall be complete and have no limitations placed on them.

-by Allen Seckinger, P.E.
Member: Probable Cause Panel

Figure 3.2 FBPE results concerning *Mary's* case.
(*Source*: Florida Board of Professional Engineers, Summer 2003 Newsletter, Florida Board of Professional Engineers, Tallahassee, FL)

3.4.5 Practicing in Your Area of Competence

Competence is an issue of relevance to engineers. Few states place any limitation upon the license itself, permitting the engineer to define his/her area of competence or expertise. The NSPE Code of Ethics, Canon 2 states that engineers should: *perform services only in areas of their competence.* A concern arises whereby an engineer seals documents that indicate electrical plans when they involve civil engineering, or structures when they are mechanical engineers by training. This type of action violates the protection of the health, safety, and welfare provision of the NSPE code of ethics and all state rules. Disciplining engineers who provide defective plans outside their area of expertise occurs frequently.

EXAMPLE 3.9

North Carolina

"D. R."

VIOLATION: Performed services outside area of competence [.0701(c)(3)]; affixed certification to inadequate design documents, failing to protect the public [.0701(b)]; produced deficient, substandard or inaccurate reports, failing to protect the public [.0701(b)]; and failed to properly certify documents [.1103].

BOARD ACTION: Reprimand and restricted the respondent from providing services as the structural engineer in responsible charge for post/pre-tensioned concrete, high risk structures (Risk Category >II; Table 1604.5, IBC 2015, as adopted by the NC Building Code), hazardous facilities, special occupancy structures, essential facilities (per Table 1604.5, IBC 2015, as adopted by the NC Building Code) over 5,000 sq. ft. in ground area or 20 ft. in height, buildings over 4 stories or 55 ft. in height, bridges with spans over 200 ft., piers with surface area greater than 10,000 sq. ft., specialty structures (e.g., communication towers, signs >100 ft. in height), and structures where 300 people or more congregate, until respondent provides proof to the Board of having successfully passed the NCEES Structural Engineering Exam (obtains acceptable results on Lateral Forces and Vertical Forces components of the exam).

EXAMPLE 3.10

Florida

"S. G."

The licensee was charged with one count of negligence when he signed, sealed, and dated plans for a force main extension that contained electrical engineering for a pump motor detail that did not comply with Rule 61G15-53.00 FAC, and with one count of practice in an area he was not qualified by training or experience to perform. After conducting an informal hearing, the Board entered into a final order imposing a $2,000 fine, cost, a reprimand of two years of probation, completion of a board-approved course in Engineering Professionalism and Ethics, and completion of the Board's Study Guide.

3.4.6 Construction Observation

During the designing of a project, it is obvious that various engineering disciplines will be involved. Likewise, during construction of a project, various subcontractors will build what was designed. To ensure the construction matches the intended design, observation of construction is required and inspection by third parties is often required.

The question is: how much observation is sufficient and by whom must it be conducted? NSPE Canon 1 suggests that the amount needed is whatever is sufficient to *hold paramount the safety, health, and welfare of the public.* Most states agree that all engineering works should pass this public health, safety, and welfare test.

While many of the prior issues provide relative consistency across all 50 states, construction observation is less uniform. Colorado provides the following language:

> *"5.3—Construction Observation as the Practice of Engineering. Section 12-25-102(10), C.R.S., defines the ". . . observation of*

> *construction to evaluate compliance with plans and specifications . . ." as the practice of engineering. Observation of construction to evaluate compliance with plans and specifications includes, but is not limited to, the following activities. (a) Observing construction operations and interpreting the project plans and specifications to monitor general compliance with the plans, specifications, and the intent of the design. (b) Evaluation or analysis of design problems due to actual field conditions encountered. (c) Evaluation or analysis of the testing of materials, equipment, or systems for acceptance, when appropriate to the project. A person who is performing, or is obligated to perform, any of the above listed activities is engaging in the practice of engineering and must either be licensed as a professional engineer in Colorado or must be supervised by a Colorado professional engineer."*

Construction defects, removal of concrete forms, incorrect rebar locations, green concrete, incorrect bolts, welds, or members, etc., are all issues that have led to failures in construction. Assigning the responsibility is a long, lengthy, legal process. But Florida has a pair of recent cases that are worth mentioning because failure to ensure compliance with the plan and specifications places the project at risk in any engineering discipline.

EXAMPLE 3.11

Florida

"L. G."

VIOLATION: Licensee was charged with violating Section 471.033(1)(g), F.S. and Rule 61G15-19.001(4), F.A.C.; negligence in the practice of engineering. Licensee was the project engineer to address site issues including lighting, drainage, utilities, landscaping, paving, signage, and other specific items. Additional services were requested. As part of the additional services, Licensee prepared and signed and sealed a Certificate of Completion which was submitted to the Building Department. Licensee certified that the wall ". . . has been constructed in substantial compliance with the permitted and approved plans." However, despite the requirement on the plans that the post caps and panel caps were to be bonded into place utilizing a silicone-based adhesive between the cap and fence component, the posts and caps were not adhered into place, remained loose, and as a result the panels and columns were never stabilized.

RULING: The case was presented to the full Board based upon a Settlement Stipulation. The Board imposed an administrative fine of $1,000, costs of $1,981.00, appearance before the Board, a reprimand, completion of a board-approved course in Engineering Professionalism and Ethics and the Board's Study Guide. A Final Order was issued on June 23, 2015. Violation: Section 471.033(1)(g), F.S. and Rule 61G15-19.001 (4), F.A.C.

EXAMPLE 3.12

Florida

"D. L."

VIOLATION: Licensee was charged with violating Section 471.033(1)(a), Florida Statutes by violating Rule 61G15-29.001, Florida Administrative Code, preparing a certification with inaccurate statements. Licensee signed, dated, and sealed a Certificate of Compliance for construction of stair railings. The Certificate stated that "To the best of my knowledge and belief, the construction of all structural load-bearing components described in the threshold inspection plan complies with the permitted documents." The stair railings were completed in a manner which was materially different than was called out on the permitted documents. There was a conflict on the drawings concerning the spacing of the rail posts. Additionally, a welded plate connection had been substituted during construction. The Certificate indicated the stair railings were a part of his threshold inspection services. The stair railings were not part of the "Structural Inspection Plan" prepared by the Engineer of Record and specifically excluded the stair railings from the required inspections. Licensee admits that the railings were not installed in accordance with the permitted drawings. Violation: Section 471.033(1)(a), Florida Statutes, and Rule 61G15-29.001, Florida Administrative Code.

RULING: This case was presented to the full Board upon a Settlement Stipulation. The Board imposed administrative costs of $1,160.75, a reprimand, appearance before the Board, successful completion of the Basic Engineering Professionalism and Ethics course and the Board's Study Guide. Final Order was issued on Aug. 15, 2019.

3.4.7 Safety

NSPE Canon 1 states: "Hold paramount the safety, health, and welfare of the public." That is pretty clear as to the obligations of engineers. Certainly, there are OSHA rules on jobsite safety, and reporting that must be complied with, and certain states cover this as well. For example, Ohio 4733-35.03 states:

> . . . *"Should the case arise where the engineer or surveyor faces a situation where the safety, health, and welfare of the public is not protected, the engineer or surveyor shall:*
>
> *(1) Sever the relationship with the employer or client;*
>
> *(2) Refuse to accept responsibility for the design, report, or statement involved;*
>
> *(3) Notify the proper authority if, in his or her opinion, the situation is sufficiently important."*

The engineer on-site has a duty, as deemed by Kant, to report a safety issue. The engineer typically has the authority to shut a jobsite down if the safety threat is imminent. If the site or job is not his, the engineer still has the responsibility to report the concern to the proper authorities. Safety is a major part of the job of an engineer.

3.4.8 Negligence—the Failure to Follow Codes

As noted in a prior section, states are beginning to require engineers to note the codes they are following on the plan sheets. Typically, building officials are the ones responsible to report engineers for failure to follow the rules. Sometimes these are egregious, sometimes just sloppy.

An infinite number of examples are available, but the key is to follow the codes. Because there are so many codes, it is probably a good idea to have another engineer review all work to ensure that the codes and designs are acceptable. Peer reviews are common in large engineering firms. Many government agencies require peer reviews and/or value engineering for large projects. The following are a few recent Florida cases with errors pertaining to codes.

EXAMPLE 3.13

Florida

"E. L."

VIOLATION: In Case No. 2017007080, Licensee was charged with violating Section 471.033(1)(g), Florida Statutes; negligence in the practice of engineering. "E. L." acted as the Structural, Electrical, Mechanical (HVAC) and Mechanical (Plumbing) Engineer of Record for the construction of a new residence. "E. L." signed, dated, and sealed Electrical Engineering documents. The documents contained material deficiencies. The deficiencies include, but are not limited to, the drawings contain an electrical riser diagram, but no short circuit values and no voltage drop calculations for the feeders and customer-owned service conductors, the conductor serving the 200 amp disconnect switch on the load side of the utility meter shows 3 No. 3/0 conductors for the phases and neutral, but no ground conductor, circuit interrupting devices are shown on the panel schedules but no reference is made regarding fault current interrupting capability, no surge protection devices are shown on the electrical drawings, the spacing of receptacles on the second plan is inadequate, etc.

"E. L." signed, dated, and sealed Mechanical (HVAC) Engineering documents. The documents contained material deficiencies. The deficiencies include, but are not limited to, the HVAC drawings contain incomplete equipment schedules for the split system to be installed, no make-up air or combustion air calculations, no ventilation design and no clothes dryer exhaust to the exterior, the drawings are inconsistent in specifying the AC condensate dry well, ductwork is shown on the drawing, but no duct is shown for outside air intake, etc.

continued

"E. L." signed, dated, and sealed Mechanical (Plumbing) Engineering documents which contained material deficiencies. The deficiencies include, but are not limited to, other than requiring that all plumbing work shall confirm to the 2014 FBC, the drawings do not state specific codes, rules, or ordinances to which the plumbing systems must comply, there is no equipment schedule to specify all plumbing fixtures, no storm water diagrams are shown, no area draining calculations are shown, etc.

"E. L." signed, dated, and sealed Structural Engineering documents which contained material deficiencies. The deficiencies include, but are not limited to, the drawing index on Sheet C1 lists a Landscape Plan and a Schedule of Windows and Doors that were not included in the initial file documentation, the drawings and specifications/notes do not provide any information regarding the design intent for roof drainage, such as the direction and magnitude of roof slope, interior roof drains versus perimeter gutters and downspouts, and the intended roofing materials and installation details, there is no information regarding the required height, materials, and details to be used for handrails and guardrails that are required, the structural calculations do not clearly specify what design dead loads were assumed in the structural design, the building elevations appear to show cantilevered concrete slabs at the west end of the second floor and roof and is inconsistent with the building sections shown on other sheets, no details were provided showing how the slabs are expected to be supported by, or connected to, the structural frame, there are no details showing typical connections between beam and slab components and the supporting columns or walls, etc.

RULING: This case was presented to the full Board upon a Settlement Stipulation. The Board imposed an administrative fine of $3,000, administrative costs of $8,500, a reprimand, appearance before the Board, a permanent restriction from practicing any electrical and mechanical engineering, probation with terms which include project review at six and 18 months, and the Board's Study Guide. Final Order was issued on 12/17/19.

EXAMPLE 3.14

Florida

"R. D."

VIOLATION: Licensee was charged with violating Section 471.033(1)(g), Florida Statutes; negligence in the practice of engineering; signing and sealing final engineering documents which contained material deficiencies. "R. D." signed, dated, and sealed structural engineering drawings for an addition which was designed as a conventionally wood framed lean-to. The documents contained material deficiencies. Those deficiencies include, but are not limited to, the wood referenced design values used by Respondent are too high and are not current, the column connection at base does not provide fixity, the headers and rafters are not designed for uplift load cases, no positive wind forces are applied to the rafters and headers in the calculations, etc.

RULING: This case was presented to the full Board upon a settlement stipulation. The Board imposed an administrative fine of $1,000, costs of $4,121.85, a reprimand, appearance before the Board, probation with terms which include successful completion of the Basic Engineering Professionalism and Ethics course, the Board's Study Guide, and project review at six and 18 months. Final Order was issued on 10/10/19.

3.5 THE MORAL OF THE STORY

The use of the term *engineer* does not necessarily provide the impression that the person is a P.E., *and* that the person is a graduate of an approved four-year engineering curriculum, as is required by the Board. *The Engineer of Record* is the person in charge who is responsible for the project, which means having a degree of control over documents, ability to make engineering decisions, responsibility to approve the work of those under him or her, and has authority over other participants. The Engineer of Record will also set expectations of *delegated engineers* and confirm that these criteria are met.

As a result, there is ultimately a series of questions that engineers must answer. For example, can the engineer knowingly:

- Practice engineering without registration?
- Use the titles "registered" or "P.E." if he or she is not one?
- Use the suffix "P.E." after his or her name if unregistered?
- Employ unlicensed people to practice engineering without a supervising P.E.?
- Give false or forged evidence to the Board or a member thereof?
- Conceal information relative to violations of applicable statutes or rules?
- Use his or her license for services, even if it is revoked or suspended?
- Seal drawings the day after his or her registration expires?

Your answer to all of these better be an emphatic *no*! Instead, engineers should:

- Exercise due care when designing, inspecting, or preparing engineering documents and providing counsel
- Practice within the areas of their true expertise
- Follow the rules!

3.6 SUMMARY

Ethics is an ongoing issue in everyone's career. In this chapter, we reviewed licensure and the expectations of engineers once licensed. Case studies were highlighted that indicate potential negligence, incompetence, and misconduct issues. Ethical violations can be any of these three, but most of the time they result in misconduct. No one wants to be labeled with the stigma of negligence, misconduct, or incompetence. The continuous goal is the protection of the public health, safety, and welfare.

Potential risks are loss of licensure, suspension and fines, etc. Every state is a little different. So, hopefully, readers have understood the origins and difficulties defining ethics. With case studies and some history from Chapter 2, the issues should become clearer. Consult your professional societies and state rules for more guidance and avoid conflicts of interest—everyone forbids these—and keep up with the changes that occur periodically within the licensing rules.

REFERENCES

Alabama Admin. Code r. 330-X-11-.03.

Alabama Law Regulating Practice of Engineering and Land Surveying. (2018). https://bels.alabama.gov/wp-content/uploads/2019/05/2019LawandCode.pdf.

Florida Administrative Code, 61G15.

FBPE.org website. Home—Florida Board of Professional Engineers (fbpe.org). Accessed 10/19/20.

Florida Statute 471. Florida Statutes > Chapter 471—Engineering. LawServer. Accessed 10/19/20. https://www.lawserver.com/law/state/florida/statutes/florida_statutes_chapter_471.

https://www.bakerdonelson.com/webfiles/Bios/50StateSurveyofLicensedDesignProfessionalStampingandSealingObligations.pdf.

https://fbpe.org/legal/disciplinary-actions/#summaries-2015-17.

https://fbpe.org/legal/statutes-and-rules/.

https://www.michigan.gov/lara/0,4601,7-154-89334_72600_72602_72731_72865---,00.html.

https://www.ncbels.org/general-info/newsletters-articles/.

https://www.sos.state.co.us/CCR/GenerateRulePdf.do?ruleVersionId=2258.

NSPE. (2020). Code of Ethics. https://www.nspe.org/resources/ethics/code-ethics. Accessed 12/27/20.

NSPE. (2020). Engineer's Creed. https://www.nspe.org/resources/ethics/code-ethics/engineers-creed. Accessed 12/27/20.

Ohio Professional Engineers and Surveyors. (2020). Rules and Regulations, Chapter 4733. https://www.peps.ohio.gov/4733/LawsandRules%7CComplete.aspx.

Rules and Regulations of the Mississippi Board of Licensure for Professional Engineers and Surveyors. (2020). https://www.pepls.ms.gov/sites/pepls/files/PEPLS%202020/Licensure%20Law/PEPLS%20Rules%20and%20Regulations%201-1-2020%20(3).pdf.

PROBLEMS

1. Explain the ethical issues involved with the following situations as they relate to the engineering profession. Identify laws, rules, canons, or other references that support your position. Provide a recommended course of action for the engineer involved in the scenario.
   a. A professional engineer is at a conference and decides he wants to play in the golf tournament. He is looking for a foursome to play. One of his college buddies, who is a major vendor that supplies water and wastewater parts to the industry, offers a spot in one of his foursomes at no charge since he is short one person anyway. The engineer is not sure if he should accept the invitation to play for free. The slot is worth $100.
   b. A professional engineer conducts asset management for a city government—including an evaluation of the sidewalks. The engineer documents the cracks and breaks in the sidewalks throughout the city. However, the city attorney instructs the engineer not to provide the information to the city because to do so would subject the city to lawsuits for negligence if people are hurt before the city can fix the cracks.
   c. Firm A is competing against other firms that have substantial political clout. Three different people approach Firm A willing to provide assistance. One says they have some negative information regarding

a couple of the competitors and are happy to get the documents to the county commission for a price. Another one offers to lobby the commission on Firm A's behalf for a portion of the contract amount. The third one employs a woman who is an acquaintance of a couple of the commissioners. They claim she can help get the job for a price.

2. A consultant is teamed with a contractor on a d/b contract, but has been asked to perform contract administration for another project where the same contractor has been engaged. What is the appropriate ethical response? Identify the appropriate licensing rule and codes of ethics to discuss this situation.
3. If an engineering firm engages a lobbyist to go after a contract, and said lobbying involves trashing the other bidders as well as perhaps overstating the firm's qualifications, what ethical issues arise? Identify the appropriate licensing rule and codes of ethics to discuss this situation.
4. Obtain the laws and rules governing professional engineers in your state and one other state. Compare and contrast the rules in a matrix.
5. Look at five years of data on cases associated with discipline in your state and one other. What are the most common issues? Make a pie chart that compares them.
6. Call or email your state's Board of Professional Engineers and ask what they perceive as the most interesting case they have had recently.
7. If you are asked to modify a specific design parameter for cost reasons, despite having performed exhaustive research that indicates only your solution will work, what should you do? Identify the appropriate licensing rule and codes of ethics to discuss this situation.

CHAPTER 4

FUNDAMENTAL EDUCATION COMPETENCY

Chapter 3 outlined licensure in states as well as common tenets that apply across them and what happens if the rules are violated. The concept of licensure stems from an engineer's responsibility to the public and the public's expectations that engineers will act to protect their interests. Most states have similar requirements to obtain a license as noted in Chapter 3, or to obtain a comity/reciprocal license. One of the typical requirements is graduation from an Accreditation Board for Engineering and Technology (ABET) accredited undergraduate program. For this chapter, the focus will turn to accreditation and the educational requirements that come with it. It is what students are in school to do—learn their profession.

LEARNING OBJECTIVES

- Gain an understanding of what accreditation is
- Gain an understanding of the history of accreditation in the United States
- Know what the requirements are for accreditation
- Know the typical requirements to meet the definition of undergraduate education
- Gain an understanding of how ethics are incorporated into education

4.1 ACCREDITATION

Ever since state engineering boards were organized in the early twentieth century, there has been a need to compare the curriculums and outcomes between colleges that offered an engineering education nationally. ABET was established in 1932 to create a structured program for engineering guidance, training, education, and recognition.

Accreditation is a nongovernmental, voluntary process involving peer review that assures that a program or institution can meet established quality standards. Students graduate with a degree that can meet certain quality requirements and that will lead to certain outcomes established by industry and society at large.

ABET is the organization that is tasked with creating these standards and then applying such toward accrediting engineering degree programs. Since its origins, accreditation via ABET has served as the basis of quality to which professional engineers (P.E.s) are held for licensure by state licensing boards. Because many states require applicants for engineering licensure to graduate from ABET-accredited undergraduate programs, nearly all engineering colleges zealously try to maintain ABET accreditation.

ABET is a nonprofit, nongovernmental organization with ISO 9001:2015 certification (ABET website, 2019). Originally, ABET was founded as the Engineers' Council for Professional Development with a mission to enhance the education of engineering professionals and students in the United States (ABET website). Within the engineering disciplines, ABET educational standards are developed from the member professional societies that commission the evaluation of degrees; the initial societies were (ABET, 2019):

- American Society of Civil Engineers (ASCE)
- American Institute of Chemical Engineers
- American Institute of Mining and Metallurgical Engineers, now the American Institute of Mining, Metallurgical, and Petroleum Engineers

- American Society of Mechanical Engineers
- American Institute of Electrical Engineers (now IEEE)
- National Council of State Boards of Engineering Examiners
- Society for the Promotion of Engineering Education, now the American Society for Engineering Education

ABET gains input from these professional societies as a means to track important knowledge that students should gain in order to ensure that graduates are current with their knowledge. The respective university departments that offer the degree programs are expected to address their comments, criticisms, and recommendations within each curriculum. ABET coordinates with the National Association of Colleges and Employers (NACE), which establishes career readiness criteria and competencies among college graduates. Table 4.1 outlines the core competencies that NACE looks for as compared to the ABET criteria.

Table 4.1 Comparison of ABET Criteria 1–7 and NACE Career Readiness Goals (*Source*: ABET)

ABET Criteria 1–7	NACE Career Readiness Scores
(1) An ability to identify, formulate, and solve complex engineering problems by applying principles of engineering, science, and mathematics	Information technology score
(2) An ability to apply engineering design to produce solutions that meet specified needs with consideration of public health, safety, and welfare, as well as global, cultural, social, environmental, and economic factors	Critical thinking/ problem solving score
(3) An ability to communicate effectively with a range of audiences	Oral/Written communication score
(4) An ability to recognize ethical and professional responsibilities in engineering situations and make informed judgments, which must consider the impact of engineering solutions in global, economic, environmental, and societal contexts	Professionalism/ work ethic score
(5) An ability to function effectively on a team whose members together provide leadership, create a collaborative and inclusive environment, establish goals, plan tasks, and meet objectives	Teamwork/ Collaboration

Continued

Table 4.1 *(continued)*

ABET Criteria 1–7	NACE Career Readiness Scores
(6) An ability to develop and conduct appropriate experimentation, analyze and interpret data, and use engineering judgment to draw conclusions	Critical thinking/ problem solving score
(7) An ability to acquire and apply new knowledge as needed, using appropriate learning strategies	Career management score

In addition, ABET establishes specific learning objectives and outcomes. Criterion 5, for example, spells out the educational requirements that ABET expects from engineering programs (ABET website, 2019).

> *"**Criterion 5. Curriculum:***
> *Curricular requirements specify topics that are appropriate to engineering technology but do not specify courses. The curriculum must combine technical, professional, and general education components in support of student outcomes. To differentiate the discipline, Program Criteria may add specificity for program curricula. The curriculum must include the following:*
>
> - ***Mathematics:** The program must develop the ability of students to apply mathematics to the solution of technical problems. Baccalaureate degree curricula will include the application of integral and differential calculus, or other mathematics above the level of algebra and trigonometry that are appropriate to the student outcomes and the discipline.*
> - ***Discipline-specific content:** The discipline-specific content of the curriculum must focus on the applied aspects of science and engineering and must:*
>   - *A. Represent at least one-third of the total credit hours for the curriculum but no more than two-thirds of the total credit hours for the curriculum;*

B. *Include a technical core preparing students for the increasingly complex technical specialties later in the curriculum;*
C. *Develop student competency in the discipline;*
D. *Include design considerations that are appropriate to the discipline and degree level, such as industry and engineering standards and codes; public safety and health; and local and global impact of engineering solutions on individuals, organizations, and society; and*
E. *Include topics that are related to professional responsibilities, ethical responsibilities, respect for diversity, and quality and continuous improvement."*

Specific disciplines may have more detailed criteria. For example, for civil engineers, ABET and ASCE, the commissioning organization, add (ABET website):

> *"Graduates of baccalaureate degree programs typically analyze and design systems, specify project methods and materials, perform cost estimates and analyses, and manage technical activities in support of civil engineering projects. The curriculum must provide instruction in the following curricular areas:*
>
> a. *Utilization of principles, hardware, and software that are appropriate to produce drawings, reports, quantity estimates, and other documents related to civil engineering;*
> b. *Performance of standardized field and laboratory tests related to civil engineering;*
> c. *Utilization of surveying methods appropriate for land measurement and/or construction layout;*
> d. *Application of fundamental computational methods and elementary analytical techniques in subdisciplines related to civil engineering;*

> e. *Planning and preparation of documents appropriate for design and construction;*
>
> f. *Performance of economic analyses and cost estimates related to design, construction, operations and maintenance of systems associated with civil engineering;*
>
> g. *Selection of appropriate engineering materials and practices; and*
>
> h. *Performance of standard analysis and design in at least three subdisciplines related to civil engineering."*

As a result, curriculums among undergraduate programs are remarkably similar in the way they attempt to meet fundamental educational competencies.

Students enrolled in an ABET-accredited engineering program are eligible to sit for the licensing exams. These exams are provided nationwide by the National Council of Examiners for Engineering and Surveying (NCEES). For examinees, the review/approval process can take up to 30 days, but will not begin until NCEES has received all required documentation.

Authorization to schedule an appointment for testing cannot be granted until this process has been completed, verified, and approved by NCEES. The Fundamentals of Engineering (FE) exam is the first step in obtaining a P.E. license. The FE exam evaluates content knowledge related to the subjects that ABET requires for accredited undergraduate engineering degree programs. Historically, the FE exam was an eight-hour exam. From 1980 to 2010, the exam evolved from individual, hand-calculated problems to an exam that contained 180 multiple-choice questions and was split evenly into a four-hour morning session (120 questions) and a four-hour afternoon session (60 questions). In 2014, the test became a 5.5-hour, computer-based exam to be administered year-round in four testing windows. The computer-based exam contains 110 multiple choice questions, within a time frame, which also includes a tutorial, a break, and a brief survey at the conclusion.

For the first half of the test, all examinees take the same general exam that is common to all engineering disciplines. Essentially, this covers information from the general education requirements and includes subjects such as: mathematics, probability and statistics, computational tools, engineering mechanics, ethics, and engineering economics. During the second half, examinees can elect to take a discipline-specific exam (chemical, civil, electrical, computer, environmental, industrial, mechanical, etc.), or a more generalized exam labeled *other disciplines.*

If the examinee holds a degree in one of the major disciplines, it is recommended that the test-taker should also sit for that discipline-specific exam. Examinees are required to participate in both sessions that they might elect on the same day.

The FE exam is closed book, but an electronic version of the FE Reference Handbook is provided on exam day. As a result, all examinees should be familiar with the FE Reference Handbook by the time they take the test. The most up-to-date version of the reference handbook is available for download or purchase at NCEES.org, along with practice exams, study materials, and other references. Some faculty members incorporate its use into the classroom to help students become familiar with using the handbook.

Candidates must bring their own calculators to the exam; however, only models of calculators as specified by NCEES can be used. The NCEES calculator policy is constantly revised because of concern for the security of examination content. Available calculator technology has been used for exam subversion, and calculators that can store and communicate text are considered a security risk. It is highly recommended to check the NCEES calculator policy for compliance and become familiar with the allowable units, so that it will not be an issue on exam day. An online computer is supplied in some test centers.

A passing score on an NCEES exam is determined by the number of correct answers or points required to indicate a knowledge level that is necessary to meet a minimal performance standard for a discipline (which is determined by an appointed committee of licensed subject-matter experts).

Beginning with the October 2005 administration, candidates received results of *pass* or *fail* only with no numerical score. There is no grading curve, but NCEES scores each exam based on its own merits and with no regard for a predetermined expectation that a certain percentage of examinees should pass or fail. All exams are scored the same way.

If a person fails the test, a diagnostic report is generated that lists the percentages of correctly answered questions in each knowledge area of the exam (see Figure 4.1). This is the best guide for determining strengths and weaknesses regarding specific subject areas.

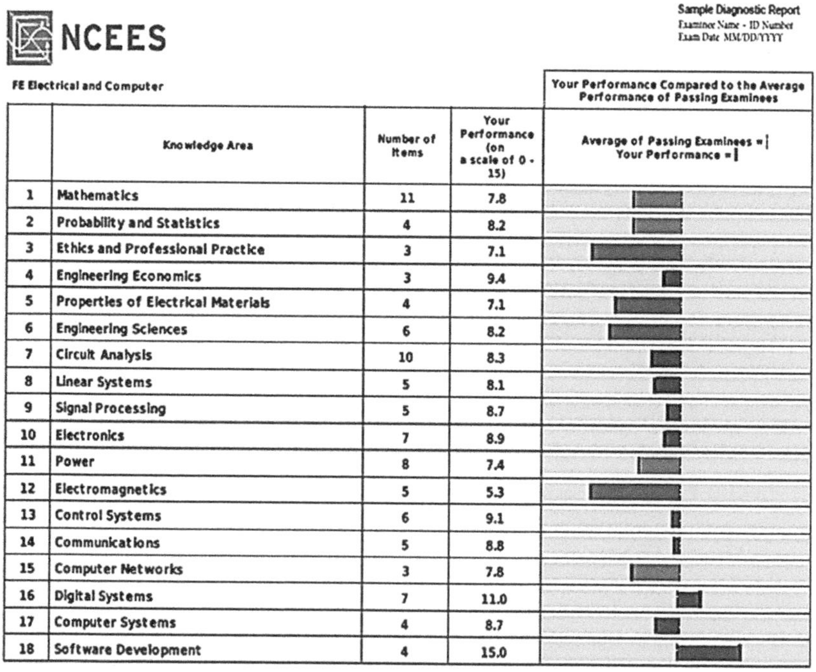

NCEES

Sample Diagnostic Report
Examinee Name - ID Number
Exam Date MM/DD/YYYY

FE Electrical and Computer

	Knowledge Area	Number of Items	Your Performance (on a scale of 0 - 15)	Your Performance Compared to the Average Performance of Passing Examinees Average of Passing Examinees = ¦ Your Performance = ▌
1	Mathematics	11	7.8	
2	Probability and Statistics	4	8.2	
3	Ethics and Professional Practice	3	7.1	
4	Engineering Economics	3	9.4	
5	Properties of Electrical Materials	4	7.1	
6	Engineering Sciences	6	8.2	
7	Circuit Analysis	10	8.3	
8	Linear Systems	5	8.1	
9	Signal Processing	5	8.7	
10	Electronics	7	8.9	
11	Power	8	7.4	
12	Electromagnetics	5	5.3	
13	Control Systems	6	9.1	
14	Communications	5	8.8	
15	Computer Networks	3	7.8	
16	Digital Systems	7	11.0	
17	Computer Systems	4	8.7	
18	Software Development	4	15.0	

Figure 4.1 Sample diagnostic report for a student *failing* the FE exam

The diagnostic report summarizes the scores as follows:

- If performance in a subject area is significantly below that of the passing examinees, this indicates that substantial study of that content area is recommended prior to retaking the exam.
- If performance in a content area is near or just above that of the passing examinees, this indicates that understanding may be improved by further study, thus improving the chances of passing the examination.

In the Figure 4.1 example, this student has significant work to do in the areas of electromagnetics, electrical circuits, materials, and engineering science—not to mention ethics. Given this is an electrical engineering student, the student has missed a significant amount of knowledge required to get licensed in his or her chosen profession in their studies to date.

For either the FE or the Principles and Practice of Engineering (PPE) exams, most states require those that fail the test more than three times enroll in and pass 12 credit hours of college classes (typically at a senior level and above) and/or an FE review course to be eligible to retake the exam.

While students who fail the test receive a report, the university also receives an institutional report as an assessment tool to help improve its curriculum. An example is shown in Figure 4.2. In this example, the university needs to improve its curriculum with respect to computational tools and ethics—both areas where the institution's students performed below the national average.

After passing the FE exam, the next step is to gain acceptable work experience, under the supervision of a licensed P.E. who would serve as a mentor. Some states have specific review classes that applicants must complete.

After gaining the appropriate experience, candidates are eligible to take the PPE exam. This is an open-book exam, but is designed to test the engineering experience gained over the four years since the candidate completed an FE exam. As a result, the PPE exam tests the

NCEES NEW REPORT FORMAT

Examination: Fundamentals of Engineering (FE)
Report title: Subject Matter Report by Major and Examination
Exams administered: January 1–May 31, 2014
Examinees included: First-Time Examinees in **EAC/ABET**-Accredited Engineering Programs
Graduation date: Examinees Testing Within 9 Months of Graduation Date

Name of Institution:		Example
Major: Civil	FE Examination:	Civil

	Institution	ABET Comparator (*2)
No. Examinees Taking (*1)	11	1632
No. Examinees Passing	10	1247
Percent Examinees Passing	91%	76%

Uncertainty Range for Scaled Score (*4) +/- 0.30

	Number of Exam Questions	Institution Average Performance Index (*3)	ABET Comparator Average Performance Index	ABET Comparator Standard Deviation	Ratio Score (*4)	Scaled Score (*4)
Mathematics	7	10.5	10.5	3.2	1.01	0.03
Probability and Statistics	4	10.8	10.9	3.8	0.99	-0.02
Computational Tools	4	9.3	10.5	3.6	0.89	-0.32
Ethics and Professional Practice	4	9.3	11.2	3.7	0.83	-0.50
Engineering Economics	4	11.9	10.4	3.7	1.15	0.42
Statics	7	10.3	10.1	2.9	1.02	0.08
Dynamics	4	12.9	10.3	3.5	1.25	0.73
Mechanics of Materials	7	10.7	9.9	2.6	1.08	0.30
Materials	4	9.8	9.4	3.0	1.05	0.15
Fluid Mechanics	4	10.7	10.9	3.5	0.98	-0.08
Hydraulics and Hydrologic Systems	8	11.0	9.4	2.2	1.17	0.75
Structural Analysis	6	10.0	9.2	2.5	1.09	0.34
Structural Design	6	9.5	8.9	2.5	1.07	0.24
Geotechnical Engineering	9	10.6	9.2	1.9	1.14	0.69
Transportation Engineering	8	9.9	9.0	2.1	1.10	0.41
Environmental Engineering	6	10.4	9.0	2.6	1.16	0.55
Construction	4	10.5	9.8	3.7	1.08	0.21
Surveying	4	10.0	8.6	3.6	1.16	0.39

Footnotes:
(*1) 0 examinees have been removed from this data because they were flagged as a random guesser.
(*2) Comparator includes all examinees from programs accredited by the ABET commission noted.
(*3) Performance index is based on a 0–15 scale.
(*4) These scores are made available for assessment purposes. See the NCEES publication entitled *Using the FE as an Outcomes Assessment Tool* at http://ncees.org/licensure/educator-resources/.

Figure 4.2 Example of a university diagnostic report

candidate's ability to competently practice in a particular engineering discipline. The PPE exam is typically the last step in the process of becoming a licensed P.E. The specific disciplines are:

- Agricultural
- Architectural
- Chemical
- Civil: Construction

- Civil: Geotechnical
- Civil: Structural
- Civil: Transportation
- Civil: Water Resources and Environmental
- Control Systems
- Electrical and Computer: Computer
- Electrical and Computer: Electrical and Electronics
- Electrical and Computer: Power
- Environmental
- Fire Protection
- Industrial
- Mechanical: HVAC and Refrigeration
- Mechanical: Mechanical Systems and Materials
- Mechanical: Thermal and Fluids Systems
- Metallurgical and Materials
- Mining and Mineral Processing
- Naval Architecture and Marine
- Nuclear
- Petroleum
- Structural I
- Structural II

The list continues to grow as the fields expand. The goal of every engineering student should be to obtain a P.E. license. The test is online as well.

4.2 EDUCATION

Currently, engineering colleges attempt to offer courses that are designed to allow graduates to be successful within a given industry. To accomplish this, advisory boards are usually developed to give input to departments and colleges concerning their programs. This includes a culminating or capstone course that allows students who are nearing graduation to put together the knowledge and skills they have acquired in their program and apply them to a major project or assignment.

In prior ABET documentation, this definition was further expanded in Criteria 5, noted previously, to include: "Students must be prepared for engineering practice through a curriculum culminating in a major design experience based on the knowledge and skills acquired in earlier course work and incorporating appropriate engineering standards and multiple realistic constraints."

Because most programs zealously guard their ABET accreditation and the input from industry, the programs have many similarities. For example, a look at a survey of 60 civil engineering programs for the 2018/2019 catalog shows the average among all schools, the range, and a comparison to ABET requirements in Criteria 5 (see Table 4.2). Column 1 contains the name of the classes. Column 2 is the average number of credits required (e.g., a "0.4" indicates many schools do not require this class) across the 60 colleges surveyed. Columns 3 and 4 define the minimum and maximum credits for a given class and Column 5 notes the ABET requirements. ABET requires 30 credits of math and science using calculus and above, including differential equations. ABET requires 45 credits of engineering classes.

Table 4.2 Typical civil engineering classes

Courses	Avg	Min	Max	ABET
Calculus 1	3.8	2.8	5	
Calculus 2	3.7	2.8	5	
Calculus 3	3.5	2.8	7.3	
Differential Equations	2.9	2	4	
Advanced Engineering Math	1.3	2	9	
Probability and Statistics	2.5	1.5	4	
Intro to Physics	0.1	0	3	
Physics 1	3.5	2.8	5	
Physics Lab	0.4	0	1	
Physics 2	2.7	0	5	
Physics Lab 2	0.4	0	1	
Physics 3	0.3	0	4	

Continued

Table 4.2 *(continued)*

Courses	Avg	Min	Max	ABET
Chemistry 1	3.4	2.1	5	
Chemistry Lab 1	0.6	0	2	
Chemistry 2	1.4	2	5	
Chemistry 3	0.1	0	3.6	
Lab	0.1	0	3	
Biology	0.5	0	4	
Geology	1.5	0	4	
Science Elective	2.0	0	9	
TOTAL MATH and SCIENCE	34.6	30	53	30
English 101	2.7	2	5	
English 102	1.6	0	4	
Technical Writing	1.3	2	3	
Humanities/Social Sciences	14.8	3	24	
Public Speaking	0.5	0	3	
American Government	0.3	0	6	
Economics	0.6	0	4	
History	0.7	0	6	
TOTAL HUMANITIES and SOCIAL SCIENCES	22.5	18	34	
Computer Programming	1.9	0	6	
Surveying	1.6	2	5	
Statics	3.0	2	6	
Dynamics	1.9	0	4	
AutoCADD	1.4	0	3	
Intro to Engineering	2.3	1	6	
Freshman Exp	0.1	0	2	
Freshman Seminar	0.0	0	2	
Civil and Environmental Engineering (CEE)	0.1	0	3	
Risk and Uncertainty	0.0	0	3	
Energy	0.0	0	3	
Reliability	0.0	0	2	
CE lab	0.0	0	2	
Soils	3.0	3	6	

Continued

Table 4.2 *(continued)*

Courses	Avg	Min	Max	ABET
Foundations	0.5	0	3	
Analysis of Structures	3.5	3	8	
Strength of Materials/Mechanics of Solids	2.7	2	6	
Civil Materials	2.8	3	6	
Reinforce Concrete	1.0	0	3	
Steel	0.7	0	3	
Hydraulics	4.2	2	10	
Hydrology	1.2	0	3	
Environmental Science	2.6	2	7	
Water/Wastewater Resources	1.0	0	6	
Thermo	1.4	0	3	
Elect. Circuits	1.4	0	5	
Economics	1.5	0	6	
Transportation 1	2.5	2	5	
Transportation 2	0.2	0	3	
Geological Eng.	0.1	0	4	
Construction	1.8	2	7	
Water Resources	0.1	0	3	
Bridges	0.0	0	3	
GIS	0.0	0	2	
Systems	0.5	0	4	
Environmental Assessment	0.0	0	2	
Subdivisions	0.0	0	0	
Sustainability	0.1	0	3	
Environmental Restoration	0.0	0	3	
Fluid Mechanics	0.0	0	2	
Tech Elective	15.0	3	44	
Building Systems	0.2	0	3	
				45
Univ. College	0.1	0	3	
Library	0.0	0	0	
Business	0.1	0	3	

Continued

Table 4.2 *(continued)*

Courses	Avg	Min	Max	ABET
Engineering Business Practice	0.1	0	3	
Professional Practice	0.3	0	2	
Professional and Legal Issues	0.2	0	3	
FE review	0.1	0	1	
Ethics	0.3	0	3	
Sr. Design Ideal including Economics and Ethics	3.9	3	7	
Unrestricted Electives	1.7	0	12	
Phys. Ed.	0.1	0	3	
TOTAL CREDITS	130	120	136	

Many states legislate the number of humanities/social science classes that students take. Typically, this number is 23 to 24 hours of humanities/social sciences. Basic classes such as statistics, dynamics, surveying, computer programming, and digital drafting consist of two or three credits, on average. Senior design, which may include ethics and economics, tends to be a two-semester class that encompasses a project that includes many aspects of civil engineering. Students work in teams to meet ABET criteria. Note that in order to create electives, many schools are less prescriptive, and as a result, data may not characterize what specific electives the students actually take. For example, materials, structural analysis, strength of materials, construction, basic transportation engineering, and environmental engineering basics are common for civil engineering students. Notice that these topics are contained in the FE exam, and thus, appear to be deemed *required learning* by most of the schools. Other programs have a similar list of classes that tie to the topics on the FE exam, whereas emerging classes do not. Examples of emerging class topics in the civil engineering programs in Table 4.2 include sustainability, risk assessment, and energy and reliability that are offered at several schools. Table 4.3 shows how the FE subjects relate directly to the classes taught. What is clear is that most schools want to maximize the success of students in civil

Table 4.3 Relationship between civil engineering courses and FE exam topics

Courses	FE Exam	Questions	FE Topic	FE Manual Sections	FE Manual Sections	FE Manual Sections
Calculus 1	x	7 to 11	Math	Algebra/ Geometry	Derivatives/ Volume	Logs/Complex Numbers
Calculus 2	x		Math	Integrals	Partial Integrals	
Calculus 3	x		Math	LaPlace	Fourier	Vectors
Differential Equations	x		Math	Differential Equations		
Adv. Engineering Math	x		Math	Matrices		
Prob. & Statistics	x	4 to 6	Math	Descrip. States	ANOVA	Distribution types
Physics	x		Science	Heat Equation	Convection	Circuits
Chemistry	x	7 to 11	Science	Periodic Table	Organic Chemistry	Structure of Matter
Computer Programming	x	4 to 6	Computers	Programming	Spreadsheets	
Surveying	x	4 to 6		Earthwork	Measurements	
Statics	x	6 to 11	Statics	Moments of Inertia	Free Body Diag.	Centroids
Dynamics	x	4 to 6	Dynamics	Kinematics	Rigid Bodies	Friction
Soils	x		Geotech	Porosity/ Saturation	Flow Nets	Stress Profiles

Continued

Table 4.3 *(continued)*

Courses	FE Exam	Questions	FE Topic	FE Manual Sections	FE Manual Sections	FE Manual Sections
Foundations	×			Bearing Capacity		
Analysis of Structures	×	7 to 11	Civil	Influence Lines	Stability/ Deflection	Loads
Mechanics of Solids	×	6 to 9	Mechanics of Materials	Torsion	Columns	Bending
Strength of Materials				Stress Type	Bending/ Torsion/Shear	
Civil Materials	×	4 to 6	Civil			
Reinforced Concrete	×	6 to 9	Civil	Internal Forces	Resistance	Shear
Steel	×		Civil	Beams	Columns	Tension of Members
Hydraulics	×	8 to 12	Civil	Fluid Flow	Power/Pumps	Flow Measurement
Hydrology	×		Civil	NCRS Method	Darcy's Law	Open Channels
Environmental Science	×	6 to 9	Civil	Streeter Phelps	Air Plumes	Mass Balance
Water/Wastewater Resources	×		Civil	Reactor Volumes	Chemical Feeds	Activated Sludge
Thermo	×		Civil	Conductance	Convection	PVRT
Elect. Circuits	×	4 to 6	Instruments	Voltage	Circuits	

Continued

Table 4.3 *(continued)*

Courses	FE Exam	Questions	FE Topic	FE Manual Sections	FE Manual Sections	FE Manual Sections
Economics	x	4 to 6	Economics	Present/ Annual Value	Depreciation	Compare Alternatives
Transportation 1	x	8 to 12	Civil	Horizontal/ Vertical Curves	Pavement Markings	Site Distance/Safety
Construction	x	4 to 6	Construction Engineering	Cost Estimates	Schedules/ CPM	Analysis of Variance
Fluid Mechanics	x	4 to 6	Fluid Mechanics	Bernouli	Statics/Liquids	
Engineering Business Practice	x					
Professional and Legal Issues	x					
Ethics	x	4 to 6	Ethics	Sustainability	Ethical Questions	Codes

engineering, so they ensure that students have the appropriate classes, as represented by the FE exam.

A survey of nearly 60 mechanical engineering programs for the 2018/2019 catalog reveals the averages in Table 4.4. As with civil engineering, ABET requires 30 credits of math and science for mechanical engineering students. Like civil students, mechanical engineering students tend to take more than the 30 credits of math and science, and 23/24 hours of humanities/social sciences.

Table 4.4 Typical mechanical engineering classes

Courses	Avg	Min	Max	ABET
Calculus 1	4.0	3	5	
Calculus 2	3.9	2	5	
Calculus 3	3.5	3	5	
Differential Equations	2.9	2	6	
Advanced Engineering Math	0.1	0	4	
Numerical Methods	0.2	0	3	
Discrete Math	0.2	0	3	
Linear Algebra	1.6	0	6	
Precalculus	0.1	0	4	
Misc. Math	1.0	0	8	
Probability and Statistics	2.3	0	4	
Intro to Physics	0.0	0	0	
Physics 1	3.8	3	5	
Physics Lab	0.3	0	2	
Physics 2	3.6	3	5	
Physics Lab 2	0.4	0	2	
Physics 3	0.9	2	7	
Chemistry 1	3.1	2	5	
Chemistry Lab 1	0.4	0	1	
Chemistry 2	0	0	4	

Continued

Table 4.4 *(continued)*

Courses	Avg	Min	Max	ABET
Biology	0.2	0	4	
Science Elective	1.4	0	7	
TOTAL MATH and SCIENCE	33.8	30	42	30
English 101	2.8	2	4	
English 102	1.6	0	6	
Technical Writing	2.2	0	5	
Humanities/Social Sciences	14.5	3	28	
Public Speaking	0.7	2	3	
American Government	0.3	0	6	
Economics	0.5	0	3	
Engineering economics	0.6	0	3	
History	0.4	0	6	
TOTAL HUMANITIES and SOCIAL SCIENCES	23.5	21	30	
Aerospace	0.2	0	6	
Applied Thermo Fluid Systems	1.3	0	11	
Digital Drafting/AutoCADD	2.1	1	6	
Circuits	2.7	0	6	
Computer Programming	3.0	1	6	
Control Systems	2.2	0	6	
Dynamic Systems	2.3	2	7	
Dynamics	2.9	2	6	
Electro-Mech. Machine design	1.8	0	6	
Electronics	0.2	0	3	
Eng. Lab	1.4	0	8	
Experimental methods	1.0	0	6	
Fabrication/Manufacturing	2.6	0	12	
Field design	0.1	0	3	
Finite elements	0.3	0	3	
Flight Mechanics	0.1	0	3	
Fluids	3.1	1	6	

Continued

Table 4.4 *(continued)*

Courses	Avg	Min	Max	ABET
Freshman Experience	0.1	0	3	
Freshman Seminar	0.2	0	4	
Gas dynamics	0.1	0	3	
Geometric modeling	0.1	0	3	
Heat Transfer	2.8	3	5	
Instrumentation	1.7	0	7	
Intro to Electrical Engineering	0.2	0	3	
Intro to Mechanical Engineering	0.6	0	7	
Intro/Foundations of Eng.	1.9	0	6	
Jr. design	0.2	0	5	
Machine design	1.9	0	6	
Materials	3.1	2	7	
Mechanical systems	0.2	0	8	
Metallurgy	0.1	0	3	
Orbital systems	0.1	0	3	
Power systems	0.1	0	6	
Product design	0.4	0	6	
Propulsion	0.1	0	3	
Sophomore design	0.5	0	5	
Statics	3.3	2	7	
Strength of Materials/MOM	3.1	3	7	
Thermodynamics	3.8	3	9	
Transport	0.3	0	3	
Vibrations	0.8	0	4	
				45
Univ. College	0.0	0	0	
Business	0.0	0	0	
Library	0.0	0	1	
Coop	0.1	0	2	
Engineering Business Practice	0	0	0	

Continued

Table 4.4 *(continued)*

Courses	Avg	Min	Max	ABET
Prof. Practice	0.4	0	4	
Professional and Legal Issues	0.0	0	1	
FE review	0.0	0	0	
Ethics	0.7	0	5	
Unrestricted Electives	3.2	0	19	
Phys. Ed.	0.2	0	4	
TOTAL CREDITS	127	120	134.5	

Typical basic classes such as statistics, dynamics, surveying, computer programming, and digital drafting are two or three credit hours each, on average. Senior design, which may include ethics and economics, tends to be a two-semester class that encompasses a project that includes all aspects of civil engineering. Students work in teams, which is part of the ABET criteria. Table 4.5 shows how the FE subjects relate directly to the mechanical engineering classes taught, and like civil degrees, the courses match up with the FE exam topics.

The same analysis is shown for electrical and computer engineering degree programs in Tables 4.6 to 4.9. Note that some electrical and computer programs are comprised into one degree with varied focus areas, much like some civil engineering degrees have an environmental focus and some mechanical degrees have an aeronautical focus. The same can be seen in other degree programs with similar results.

Table 4.5 Relationship between mechanical engineering courses and FE exam topics

Courses	FE Exam	Questions	FE Topic	FE Manual Sections	FE Manual Sections	FE Manual Sections	FE Manual Sections
Calculus 1	x	11 to 17	Algebra/ Geometry	Derivatives/ Volume	Logs/ Complex Numbers		
Calculus 2	x		Integrals	Partial Integrals			
Calculus 3	x		LaPlace	Fourier	Vectors		
Differential Equations	x		Differential Equations				
Advanced Engineering Math	x		Matrices				
Probability and Statistics	x	4 to 6	Descrip. States	ANOVA	Distribution Types	Linear Regression	
Physics	x		Heat Equation	Convection	Circuits	Instruments	
Chemistry	x	7 to 11	Periodic Table	Organic Chemistry	Structure of Matter	Mass Transfer	
Computer Programming	x	4 to 6	Programming	MATLAB	C++		
Electrical Materials	x	4 to 6	Chemical	Corrosion	Piesometric	Thermal	

Continued

Table 4.5 *(continued)*

Courses	FE Exam	Questions	FE Topic	FE Manual Sections	FE Manual Sections	FE Manual Sections	FE Manual Sections
Engineering Science	x	6 to 8	Work	Voltage	Power	Conductors	Capacitance
Circuits Analysis	x	10 to 15	KCL/KVL	Series/Loop Circuits	Norton/ Thevian	Node/Loop Analysis	Waveform
Linear Systems	x	5 to 8	Frequency/ Transient	Resonance	LaPlace	Transfer Function	2 Port
Signal Processing	x	5 to 8	Convolution	Differential Equations	Z Transforms	Nyquist	Analog Filters
Electronics	x	7 to 11	Solid State	Discrete	Amplifiers	Instrumentation	Power Electronics
Power Systems	x	8 to 12	Power Factors	Motors and Generators	Transmission and Distribution	Transformers	Single Phase
Electromagnetics	x	5 to 8	Maxwell Eq.	Electrostatics	Wave Propagation	Transmission Lines	Emag Compatibility
Control Systems	x	6 to 8	State Variables	Controllers	Closed/Open Loop	Bode Plots	Block Diagrams
Communications	x	5 to 8	Base Modulations	Fourier Tranforms	Multiplex	Digital Comm.	
Computer Networks	x	3 to 5	Rotating and Switching	Network Topologies	Frameworks	LANs	

Continued

Table 4.5 *(continued)*

Courses	FE Exam	Questions	FE Topic	FE Manual Sections	FE Manual Sections	FE Manual Sections	FE Manual Sections
Digital Systems	x	7 to 11	Architecture	Microprocessors	Memory Tech	Interfacing	
Software Development	x	4 to 8	Software Testing	Software Design	Data Structures	Algorithms	
Economics (Sr Design)	x	3 to 6	Economics	Present/Annual Value	Depreciation	Compare Alternatives	
Engineering Business Practice (Design 2)	x						
Ethics (Design 1)	x		Sustainability	Ethical Questions	Codes	Societal Objectives	
Fluid Mechanics	x	4 to 6	Bernoulli	Statics/Liquids			
Engineering Business Practice	x						
Professional and Legal Issues	x						
Ethics	x	4 to 6	Sustainability	Ethical Questions	Codes	Societal Objectives	

Table 4.6 Typical electrical engineering classes

Courses	Avg	Min	Max	ABET
Calculus 1	4.0	3	5	
Calculus 2	3.9	2	5	
Calculus 3	3.5	3	5	
Differential Equations	2.9	2	6	
Adv. Engineering Math	0.1	0	4	
Numerical Methods	0.2	0	3	
Discrete Math	0.2	3	3	
Linear Algebra	1.6	0	6	
Precalculus	0.1	0	4	
Misc. Math	1.0	2	8	
Prob. & Statistics	2.3	0	4	
Intro to Physics	0.0	0	0	
Physics 1	3.8	3	5	
Physics Lab	0.3	0	1	
Physics 2	3.6	3	5	
Physics Lab 2	0.4	0	1	
Physics 3	0.9	2	7	
Chemistry 1	3.1	2	5	
Chem. Lab 1	0.4	0	1	
Chemistry 2	0.1	0	4	
Biology	0.2	3	4	
Science Elective	1.4	0	6	
TOTAL MATH and SCIENCE	34	30	42	30
English 101	2.8	2	4	
English 102	1.6	0	6	
Technical Writing	2.2	1	5	
Humanities/Social Sci.	14.5	3	24	
Public Speaking	0.7	2	3	
American Govt.	0.3	0	6	
Economics	0.5	0	3	
Engineering Economics	0.6	0	3	
History	0.4	0	6	
TOTAL HUMANITIES and SOCIAL SCIENCES	23.5	21	34	

Continued

Table 4.6 *(continued)*

Courses	Avg	Min	Max	ABET
Analog Systems	1.0	0	6	
Circuits	5.5	3	12	
Communication Systems	0.8	0	4	
Communications	0.0	0	0	
Computer Programming	3.1	1	18	
Control Systems/Instr.	0.8	0	6	
Data Structures	0.3	3	4	
Design of Eng. Systems	0.3	0	3	
Digital Logic	2.2	1	6	
Digital Systems	0.6	0	8	
Discrete Systems	0.2	0	4	
EE Lab	0.7	0	5	
Electr. Machines/Devices	0.7	0	8	
Electric Power/Transmission	1.3	2	8	
Electromagnetic waves	3.1	0	6	
Embedded Systems	0.8	0	4	
Eng. Design Think	0.8	0	6	
Engineering Analysis	0.5	0	6	
Foundations of EE	0.0	0	0	
Foundations of Computer Eng./EE	0.4	0	4	
Freshman Exp.	0.1	0	1	
Freshman Seminar	0.0	0	2	
Intro to Stoch Sus.	0.0	0	0	
Intro to Computer Organ.	0.8	3	4	
Intro to CS	0.7	2	6	
Intro to EE	0.8	1	4	
Intro to Eng. Applied Sci.	0.1	0	4	
Intro to Eng.	1.3	0	8	
Jr. Project	0.3	0	4	
Linear Systems	1.1	0	6	
Materials	0.2	0	4	
Microprocessors	1.9	0	12	

Continued

Table 4.6 *(continued)*

Courses	Avg	Min	Max	ABET
Nano-systems	0.0	0	0	
Networks	0.4	3	4	
Optics	0.1	0	3	
PLCs	0.1	0	4	
Project Management	0.1	0	3	
Semiconductor/Electronics	3.5	3	11	
Signal Processing	0.6	0	5	
Signals and Systems	2.7	3	8	
So. Project	0.1	0	3	
Software Design	0.5	0	4	
Solid State	0.1	0	3	
Statics/Strengths	0.4	0	4	
Stoich/Random Signals	0.7	0	3	
Thermodynamics	0.2	0	4	
Troubleshooting Electrical Systems	0.1	0	3	
Sr. Design/Capstone	4.7	3	7	
Technical Electives	20.6	6	40	
				54
Univ. College	0.0	0	0	
Business	0.0	0	0	
Library	0.0	0	1	
Coop	0.1	0	2	
Engineering Business Practice	0.0	0	0	
Prof. Practice	0.4	0	3	
Prof. & Legal Issues	0.0	0	1	
FE review	0.0	0	1	
Ethics	0.7	0	3	
Unrestricted Electives	3.2	0	15	
Phys. Ed.	0.2	0	3	
TOTAL CREDITS	122.2	120	136	

Table 4.7 Relationship between electrical engineering courses and FE exam topics

Courses	FE Exam	Questions	FE Topic	FE Manual Sections	FE Manual Sections	FE Manual Sections	FE Manual Sections
Calculus 1	x	11 to 17	Algebra/ Geometry	Derivatives/ Volume	Logs/ Complex Numbers		
Calculus 2	x		Integrals	Partial Integrals			
Calculus 3	x		LaPlace	Fourier	Vectors		
Differential Equations	x		Differential Equations				
Advanced Engineering Math	x		Matrices				
Probability and Statistics	x	4 to 6	Descrip. States	ANOVA	Distribution Types	Linear Regression	
Physics	x		Heat Equation	Convection	Circuits	Instruments	
Chemistry	x	7 to 11	Periodic Table	Organic Chemistry	Structure of Matter	Mass Transfer	
Computer Programming	x	4 to 6	Programming	MATLAB	C++		
Electrical Materials	x	4 to 6	Chemical	Corrosion	Piesometric	Thermal	
Engineering Science	x	6 to 8	Work	Voltage	Power	Conductors	Capacitance

Continued

Table 4.7 *(continued)*

Courses	FE Exam	Questions	FE Topic	FE Manual Sections	FE Manual Sections	FE Manual Sections	FE Manual Sections
Circuits Analysis	x	10 to 15	KCL/KVL	Series/Loop Circuits	Norton/ Thevian	Node/Loop Analysis	Waveform
Linear Systems	x	5 to 8	Frequency/ Transient	Resonance	LaPlace	Transfer Function	2 Port
Signal Processing	x	5 to 8	Convolution	Differential Equations	Z Transforms	Nyquist	Analog Filters
Electronics	x	7 to 11	Solid State	Discrete	Amplifiers	Instrumentation	Power Electronics
Power Systems	x	8 to 12	Power Factors	Motors and Generators	Transmission and Distribution	Transformers	Single Phase
Electromagnetics	x	5 to 8	Maxwell Eq.	Electrostatics	Wave Propagation	Transmission Lines	Emag Compatibility
Control Systems	x	6 to 8	State Variables	Controllers	Closed/Open Loop	Bode Plots	Block Diagrams
Communications	x	5 to 8	Base Modulations	Fourier Tranforms	Multiplex	Digital Comm.	
Computer Networks	x	3 to 5	Rotating and Switching	Network Topologies	Frameworks	LANs	

Continued

Table 4.7 *(continued)*

Courses	FE Exam	Questions	FE Topic	FE Manual Sections	FE Manual Sections	FE Manual Sections	FE Manual Sections
Digital Systems	x	7 to 11	Architecture	Microprocessors	Memory Tech	Interfacing	
Software Development	x	4 to 8	Software Testing	Software Design	Data Structures	Algorithms	
Economics (Sr. Design)	x	3 to 6	Economics	Present/Annual Value	Depreciation	Compare Alternatives	
Engineering Business Practice (Design 2)	x						
Ethics (Design 1)	x		Sustainability	Ethical Questions	Codes	Societal Objectives	
Fluid Mechanics	x	4 to 6	Bernoulli	Statics/Liquids			
Engineering Business Practice	x						
Professional and Legal Issues	x						
Ethics	x	4 to 6	Sustainability	Ethical Questions	Codes	Societal Objectives	

Table 4.8 Typical computer engineering classes

Courses	Avg	Min	Max	ABET
Calculus 1	3.9	3	5	
Calculus 2	3.8	2	5	
Calculus 3	2.9	3	5	
Differential Equations	2.4	2	4	
Advanced Engineering Math	0.1	0	3	
Discrete Math	1.1	0	4	
Linear Algebra	1.5	2	4	
Matrices	0.2	0	3	
Misc. Math/Numerical Methods	0.9	0	6	
Probability and Statistics	2.5	2	7	
Intro to Physics	0.0	0	0	
Physics 1	3.9	3	5	
Physics Lab	0.3	0	1	
Physics 2	3.8	3	5	
Physics Lab 2	0.4	0	5	
Physics 3	0.6	0	7	
Chemistry 1	2.8	0	5	
Chem. Lab 1	0.3	0	1	
Chemistry 2	0.2	0	3	
Chemistry 3	0.0	0	0	
Lab	0.0	0	0	
Biology	0.2	0	4	
Geology	0.0	0	0	
Science Elective	1.3	0	7	
TOTAL MATH and SCIENCE	33.2	30	43	30
English 101	2.9	2	4	
English 102	1.8	0	6	
Technical Writing	2.1	0	6	
Humanities/Social Sciences	15.1	3	24	
Public Speaking	0.9	0	4	

Continued

Table 4.8 *(continued)*

Courses	**Avg**	**Min**	**Max**	**ABET**
American Government	0.3	0	6	
Economics	0.2	0	3	
Engineering Economics	0.4	0	3	
History	0.5	0	6	
TOTAL HUMANITIES and SOCIAL SCIENCES	24.2	21	33	
Computer Programming	5.4	0	17	
Circuits	5.1	2	11	
Logic Design	2.0	3	9	
Signals and Systems	2.4	3	8	
Data Structures	2.5	3	5	
How the Internet Works	0.1	0	0	
Intro to Eng.	1.3	0	8	
Engineering Design/Think	0.2	0	4	
Freshman Experience	0.2	0	3	
Freshman Seminar	0.1	0	1	
Digital Drafting/AutoCADD	0.1	0	0	
Advanced Computing	0.1	0	4	
Algorithms	0.9	0	4	
Analog Systems	0.4	0	6	
Communication Systems	0.2	0	4	
Computer Architecture	1.5	0	4	
Computer Design	0.1	0	3	
Computer System Performance	0.1	0	3	
Design of Computer Systems	0.5	0	4	
Digital Processing	1.3	0	5	
Discrete Computer Design	0.9	0	7	
Discrete Systems	0.8	0	4	
Drafting	0.0	0	0	
Electrodynamics	0.2	0	3	
Electromagnetics	0.3	0	4	

Continued

Table 4.8 *(continued)*

Courses	Avg	Min	Max	ABET
Electronics	2.2	0	8	
Embedded Systems	1.3	0	4	
Engineering Analysis	0.2	0	3	
Foundation of Computer Algorithms	0.4	0	8	
Internet of Things	0.0	0	1	
Intro to Computer Engineering	0.4	0	6	
Intro to Computer Science	0.6	0	7	
Intro to Computer Systems	1.0	1	4	
Intro to ECE	0.4	0	7	
Intro to EE	0.3	0	4	
Library Instruction	0.0	0	0	
Linear Systems	0.5	0	6	
Microprocessors/Digital Systems	1.7	0	8	
Networks/Computer Organ.	2.0	0	13	
Operating Systems	1.6	0	8	
Programming Systems	0.8	0	4	
Random Systems	0.4	0	4	
Robotics	0.2	0	8	
Electromagnetic Fields	0.1	0	4	
Security	0.2	0	3	
Software Design	1.5	0	9	
Statics	0.1	0	3	
Stochastic Models	0.0	0	0	

Continued

Table 4.8 *(continued)*

Courses	Avg	Min	Max	ABET
Telecom	0.2	0	3	
Thermodynamics	0.2	0	6	
Unix	0.1	0	1	
VLS Design	0.1	0	4	
Web Design	0.1	0	3	
Database Management	0.1	0	3	
Lab	0.4	0	0	
So. Seminar	0.1	0	3	
Sr. Design	4.4	2	8	
Tech Electives	17.2	3	33	
				45
Univ. College	0.0	0	1	
Business	0.1	0	3	
Jr. Seminar	0.2	0	1	
Engineering Business Practice	0.0	0	1	
Professional Practice	0.2	0	3	
Professional and Legal Issues	0.0	0	1	
FE Review	0.0	0	0	
Coop	0.1	0	2	
Ethics I	0.1	0	3	
Ethics II	0.8	0	4	
Unrestricted Electives	2.7	0	15	
Phys. Ed.	0.2	0	3	
TOTAL CREDITS	127	120	132	

Table 4.9 Relationship between computer engineering courses and FE exam topics

Courses	FE Exam	Questions	FE Topic	FE Manual Sections	FE Manual Sections	FE Manual Sections	FE Manual Sections	FE Manual Sections	FE Manual Sections
Calculus 1	x	11 to 17	Math	Algebra/ Geometry	Derivatives/ Volume	Logs/Complex Numbers			
Calculus 2	x		Math	Integrals	Partial Integrals				
Calculus 3	x		Math	LaPlace	Fourier				
Differential Equations	x		Math	Differential Equations	Vectors				
Advanced Engineering Math	x		Math	Matrices					
Probability and Statistics	x	4 to 6	Math	Descrip. States	ANOVA	Distribution Types	Linear Regression		
Physics	x		Science	Heat Equation	Convection	Circuits	Instruments		
Chemistry	x	7 to 11	Science	Periodic Table	Organic Chemistry	Structure of Matter	Mass Transfer		
Computer Programming	x	4 to 6	Computers	Programming	Spreadsheets				
Electrical Materials	x	4 to 6	Chemical	Corrosion	Piesometric	Thermal			

Continued

Table 4.9 *(continued)*

Courses	FE Exam	Questions	FE Topic	FE Manual Sections	FE Manual Sections	FE Manual Sections	FE Manual Sections	FE Manual Sections	FE Manual Sections
Engineering Science	x	6 to 8	Work	Voltage	Power	Conductors	Capacitance	Electrics Fields	Inductance
Circuits Analysis	x	10 to 15	KCL/KVL	Series/Loop Circuits	Norton/ Thevian	Node/Loop Analysis	Waveform	Phasors	Impedence
Linear Systems	x	5 to 8	Frequency/ Transient	Resonance	LaPlace	Transfer Function	2 Port		
Signal Processing	x	5 to 8	Convolution	Differential Equations	Z Transforms	Nyquist	Analog Filters	Digital Filters	
Electronics	x	7 to 11	Solid State	Discrete	Amplifiers	Instrumentation	Power Electronics		
Power Systems	x	8 to 12	Power Factors	Motors and Generators	Transmission and Distribution	Transformers	Single Phase		
Electromagnetics	x	5 to 8	Maxwell Eq.	Electrostatics	Wave Propagation	Transmission Lines	Emag Compatibility		
Control Systems	x	6 to 8	State Variables	Controllers	Closed/Open Loop	Bode Plots	Block Diagrams		
Communications	x	5 to 8	Base Modulations	Fourier Transforms	Multiplex	Digital Comm.			

Continued

Table 4.9 *(continued)*

Courses	FE Exam	Questions	FE Topic	FE Manual Sections	FE Manual Sections	FE Manual Sections	FE Manual Sections	FE Manual Sections	FE Manual Sections
Computer Networks	x	3 to 5	Rotating and Switching	Network Topologies	Frameworks	LANs			
Digital Systems	x	7 to 11	Architecture	Micro-processors	Memory Tech	Interfacing			
Software Development	x	4 to 8	Software Testing	Software Design	Data Structures	Algorithms			
Economics	x	3 to 6	Economics	Present/ Annual Value	Depreciation	Compare Alternatives			
Engineering Business Practice	x								
Professional and Legal Issues	x								
Ethics	x	3 to 6	Ethics	Sustainability	Ethical Questions	Codes	Societal Objectives		

Ultimately, the conclusion one might reach is that many engineering schools offer pretty much the same classes for similar degree programs. The focus may vary depending on local industry needs (mechanical engineering at Wayne State University in Detroit, Michigan, for example, focuses on automobiles because of its location. By comparison, Florida Atlantic University in Boca Raton, Florida, serves a community that has no local car manufacturing industry but produces renewable power and aircraft engines).

Consistency also leads to accreditation (something that all schools judiciously defend) and is likely to be maintained. The only problem with consistency is when there are changes to local or national industries. In such cases, the change may be slow or, as many schools have found, the new discipline (computer engineering, for example) may come from an existing, accredited program (like electrical engineering).

4.3 TEACHING ETHICS IN COLLEGE

The main engineering ethics problem that college students face is a lack of academic integrity. A lack of academic integrity can show itself in the form of cheating by copying someone's work, intentional test cheating, plagiarism, and/or self-plagiarism (Nyugen, 2013). While completely inappropriate and grounds for dismissal in most colleges, it is not life threatening. However, as students transition to their careers, professional ethics rises in importance. As a result, while engineering curricula has evolved, the academic accrediting bodies such as ABET now require ethics to be formally taught in colleges and universities (ABET, 2015). In 1985, ABET decided to require engineering programs that wanted to be ABET accredited to provide students with "an understanding of the ethical characteristics of the engineering profession and practice." In 2000, ABET approved even more specific requirements assuring that engineering graduates have not only an understanding of ethical and professional issues related to the practice of engineering, but also an understanding of the impact of engineering on larger social issues.

Ethics is a significant component of the FE exam provided by NCEES. However, several challenges to ethics education remain (as indicated in the decreasing scores reported in Chapter 1). The first is that ethics is, by nature, often ambiguous, which is not how engineers are taught to think. Engineers want answers. As a result, situations are conceptually difficult for people to understand and assess.

In addition, most engineering faculty members typically lack real-world experience. As a result, they have never been faced with complex ethical dilemmas that can be encountered only in professional practice (Marcy and Rathburn, n.d.). This lack of practical experience is often coupled with a lack of willingness by faculty to teach the subject or to deal with the abstract philosophical concepts associated with engineering ethics. You cannot teach what you don't know.

Because engineers must recognize the importance of compliance with their legal and ethical obligations, Florman (1983) suggested five goals to address within academic areas:

1. To stimulate the moral imagination of students
2. To help students recognize ethical issues
3. To help students analyze key moral concepts and principles
4. To stimulate a sense of moral responsibility
5. To help students deal constructively with moral ambiguity and disagreement

Analyses of case studies have proven one of the most popular and effective ways of pursuing these goals.

Supplying still more impetus to the development of engineering ethics, it has been revealed that most U.S. engineers do not have the P.E. license, which means that the penalties for failure to follow ethical principles in their engineering work lack any real punishment. Meanwhile, licensed P.E.s continually face the possibility of losing their license if they do engage in unethical conduct.

4.4 SUMMARY

Accreditation is a means that licensing boards use to ensure consistency, value, and competency within engineering education. ABET is that accreditation board. ABET sets standards that it attempts to quantify and validate when visiting universities in order to determine whether the program should retain (or obtain) accreditation. Understanding what the ABET standards are and what reviewers seek in a standardized engineering program results in a situation where most programs are similar in what they offer in terms of curriculum. The challenge comes when universities must quickly react as industrial needs continually change.

REFERENCES

ABET. (2015). *Accrediting Board for Engineering and Technology.* http://www.abet.org/.

ABET 2020. (2019–2020). *Criteria for Accrediting Engineering Programs, 2019–2020.* https://www.abet.org/accreditation/accreditation-criteria/criteria-for-accrediting-engineering-programs-2019-2020/.

Florman, Samuel. (1983). Commentary. In *The DC-10.* John H. Fielder and Douglas Birsch (eds.). State University of New York Press. Albany, NY.

Marcy, W.M. and J.B. Rathbun. (n.d.). *Engineering Ethics and Its Impact on Society.* https://ethicalengineer.ttu.edu/articles/engineering-ethics-and-its-impact-on-society. Accessed 2/14/2021.

Nyugen, D. (2013). *Engineering Ethics.* https://sites.tufts.edu/eeseniordesignhandbook/2013/engineering-ethics-2/. Accessed 10/14/2020.

PROBLEMS

1. Obtain five profiles from schools that are ABET accredited and compare your course offerings with theirs. Then compare them to ABET. What is missing?
2. Course match your program with the NCEES subjects tested. Outline the courses in which each test subject is covered.

This book has free material available for download from the Web Added Value™ resource center at *www.jrosspub.com*

CHAPTER 5

WORKING AS AN ENGINEER

The goal of this chapter is to explain how engineers procure work, what clients seek, and how engineers function. State laws may guide the selection of engineers. For example, some states prohibit the use of cost as a category for evaluating engineering projects—preferring to focus on credentials. The rationale is that those who are better qualified to do the work will do it more efficiently and with less risk. Most engineering societies approve of this selection criteria because it puts the best qualified engineers—those least likely to make errors or struggle with design—in control of the project. That meets the public interest test as well as the goals of codes of ethics that were created for engineers.

Typically, the private sector is free to do as it wishes. It can use price or other means as a criterion to select consultants, although many private organizations are leaning toward public sector models. In effect, they are attempting to find the best qualified firms to efficiently undertake a project versus contracting a firm that is less costly but may not be able to perform the task as well.

In this chapter, the topic will be procuring work through a proposal process. The type of organization may matter—whether it be individuals or corporations. Once engineers have been selected to begin a project, bringing in others who are critically needed to finalize contracts is the next challenge. Work will progress, but communication remains the key to success. Ethics is how engineers show their past work, seek new work, and help complete it once work is obtained.

LEARNING OBJECTIVES

- Identify what clients seek from consultants
- Describe the proposal process regarding what is needed and the typical items that must be included
- Explain the practice of working as an engineer on a project (of any type)
- Define the types of organizations that engineers may create in order to produce results—and why

5.1 WHAT CLIENTS SEEK IN THE SELECTION OF ENGINEERS

Protection of the public health, safety, and welfare is the primary public trust responsibility for all engineers. As a result, obtaining work, especially in the public sector, is the first action in which an engineer can encounter trouble. Most owners generally want the best qualified people to provide the engineering services required (although they may not like the price). Some engineers will submit proposals for work and then actually lobby elected officials to obtain that work, even though they have no prior experience to do that work. That practice not only violates the honesty and forthrightness criteria, but such an action also violates the ethical requirement to perform work only in areas of competence. If one has no expertise, it is hard to argue your competency to do the work. Lobbying to get work might involve damaging the reputation of other competing engineers to gain leverage. This clearly violates the universal ethical goal to protect the profession, a specific canon in most codes of ethics. Since engineering relies on public trust, disparaging other engineers is a clear violation of this ethical principle.

5.1.1 What Clients (Should) Seek from Consulting Engineers

What clients/stakeholders establish as criteria when selecting a consultant is not always what might be expected. Most clients look for honesty, dependability, and forthrightness. But these criteria are more subjective than literal. Clients seek consultants that will be truthful, indicate what problems might exist, and continue to be candid about issues rather than furtive (Bloetscher, 1999 and 1999a). Because clients generally seek consultants who are best qualified to provide the type of work required, trust is of great importance to the selection process (Bloetscher, 1999 and 1999a).

Not all consultants are equally qualified, despite any given consultant's self-perception. For instance, if an owner wants to build a membrane water treatment plant, a consultant who has already designed a similarly sized membrane facility would be preferred. Yet, despite this obvious consideration, consultants without experience will submit on the work and some will actually lobby elected officials in an attempt to obtain it (Bloetscher, 1999 and 1999a). This violates the honesty and forthrightness criteria, and as such, is certainly an ethical violation as well. It is far better for consultants who do not have any specific type of experience to team up with consultants who do. Thereby, any lobbying efforts are maintained within more appropriate parameters, although the very act of lobbying itself may still constitute an ethical or legal problem (Bloetscher, 1999 and 1999a).

Putting the considerations of the honesty of the consultant aside, an employer must attempt to procure the right skills to do the job. This not only measures the known capabilities of the firm, but also serves as a measurement of the specific talents of the personnel who will be selected by the organization to perform the work.

A common problem is that consultants can submit proposals that represent the capabilities of people who will never actually work on the proposed project. All of this is done in an attempt to show that the firm has the expertise to do a job. This tactic of enhancement is misleading and inherently dishonest. This serves as evidence that the consultant is not putting sufficient thought into the proposal process, which is likely to end up with a poor response from the selection committee. It is also a violation of most codes of ethics—regarding both the part on dishonesty and the requirement for practicing in your area of competence only.

Accessibility of the consultant is also an important consideration when performing any task. However, this does not mean that consultants need to have an office within the community in which the job is located, as has been demonstrated in the COVID-19 world through the popular acceptance of Zoom, Webex, GoToMeeting, and many other online meeting platforms that came of age in 2020. If an important member of the consulting team can be assigned to a jobsite who will be available to attend a meeting within two to four hours or be available by phone, Zoom, or other method at a moment's notice, this will usually meet the *accessibility* criteria. For a complex water treatment plant, the owner should not expect that the consultant would be able to solely rely upon local staff to do *all* of the design work.

A favorable reputation among other public sector bodies (and within the consulting field in general) is extremely important to the success and continued growth of an engineering consulting firm. Therefore, any references should include those from other consultants as well as from other governing bodies.

The world within the engineering field is surprisingly small, and impressions and knowledge of issues that arise between individual engineers (and contractors) spreads quickly. A poor reputation in relationship to working with other consultants reflects negatively on the firm and its employees.

Another important criterion is performance, particularly when trying to meet time and budgetary constraints. Past schedule and project budget performance will indicate whether the consultant will apply the appropriate staff and resources to complete the work in a timely manner. It is indicative of the consultant's ability to be flexible in meeting the needs of the project while maintaining progress toward timeliness and budget.

Finally, the owner should determine whether or not the people assigned to the job as project managers (PMs) are comfortable in dealing with the consultants. This goes back to honesty and forthrightness. If the client's personnel are not comfortable, or do not trust the consultant's staff, the relationship will be strained going forward.

5.1.2 What Consultants Do Not Need

There are often issues that consultants think they need to address, but actually do not, from a staff perspective. Consultants do not need to represent the largest national or international consulting firms. Big national firms may be able to access any number of experts to provide information, but their focus cannot be with one client, especially smaller ones. As a result, a more local consultant who has access or collaborative efforts involving small, local providers matched with a larger consulting firm may be seen as a more palatable alternative.

A single consultant does not need to personally provide all of the skills to do the job but must be able to take responsibility for the important ones. If the consultant can bring in content experts from other offices or tap specific skill sets from among the team members, it will strengthen the quality of the proposal. In any case, the abilities of the other members of the consulting team should address the missing skills of the primary consulting firm. It is also helpful if members of the team have experience in working together as an effective team. Clients will typically attempt to identify the firm with the people who are most capable of performing the job and are wary of using public monies or the public trust to be the first on the block to try something new.

5.2 THE PROPOSAL PROCESS

In most cases, consultants are chosen based on qualifications, not price. There are exceptions, but they are not the norm—and for good reason. Low price and competence are not always related. Complex projects cost more to design. Assuming a qualification-based approach, the process to secure a project should include one of the following:

- **Request for qualifications (RFQ):** The owner advertises a proposed project and asks that consulting firms submit their experience and testimonials.

- **Statement of interest and qualifications (SIQ):** The consultant's response to an SIQ is typically to submit a letter expressing interest in a particular project along with a package outlining the team's qualifications.
- **Request for proposals (RFP):** The owner advertises a proposed project and asks that consulting firms submit their proposed ideas to complete the assignment, their qualifications to do the work, and often a proposed budgetary outline. From an engineering perspective, this is most often used for projects where the selection involves a team that will design the project and manage its construction.

Each of the approaches involve roughly the same requirements. One important aspect of the SIQ/RFQ process is understanding the approach to the project and the management of the same. The following paragraphs outline how such a request might appear (based on an actual template).

5.2.1 Scope of the Project

In the original SIQ, RFQ, or RFP package, the scope of the project may or may not be well-defined. It is up to the consultants to refine the scope of the project because the original authors of the request may not have a clear understanding of what might be feasible to meet the needs of the project. The user typically provides a rough outline for the scope of services. To demonstrate a complete understanding of the project to the client, an effective response explains the objectives of the project in much more detail.

For example, the following is a typical scope that represents a poorly written and misdefined example of the effort contemplated.

EXAMPLE 5.1

A Poorly Defined Scope

The city desires to retain a design-build project team to provide the scope of services described as follows:

1. Land surveying
2. Site planning
3. Construction of the improvements designed
4. Various studies, reports, etc., as necessary to accomplish the work
5. Decommission and removal of existing fuel station within six months

This scope of services may be expanded or reduced at the discretion of the city to either include or remove any one or more service.

Compare the scope in Example 5.1 to a better-defined scope in Example 5.2. Which one will provide a better understanding of the project and result in a better opportunity to demonstrate qualifications to pursue the project? If selected, the engineer will be pricing the work based on the scope, so a better-defined scope will improve pricing accuracy. The more defined the scope, the less chance there is for misunderstanding, a major component of risk in any project.

EXAMPLE 5.2

A Better-Defined Scope

The city has created a Transit Oriented Development (TOD) zone in its downtown area as a part of a Regional Activity Center. At present, the downtown area is blighted (as a result of economic conditions) with older, mostly vacant buildings that are not conducive to an area that is likely to use alternative modes of transportation. The desire of the

continued

city is to redevelop the downtown commercial district to create employment opportunities and an increased tax base. Ultimately the city wants 650,000 square feet of commercial space downtown and will add 4,000 residential mixed-use housing units. The city is looking to expand its ocean access and boating industry and desires that the downtown concept be in keeping with this desire. The theme is that the city is a "Marine Town" and, therefore, its redevelopment and architecture should be in keeping with that theme.

At the present time the city is constructing a Leadership in Energy and Environmental Design (LEED®)-certified branch of the County Public Library at its current City Hall site. The city also is evaluating a 400-car parking garage on the site just south of City Hall. The city also owns the three lots directly north of the City Hall site, and desires to convert them to a mixed-use development with all parking on-site. It is desired to have this location incorporate 30,000 square feet of commercial space and twenty, 2 bedroom/2 bath (minimum) condo units. The building will need to be LEED® Gold, meet all TOD zone criteria, and hold all of its stormwater on-site.

To use this site, the city will need to have the fire station moved to a better location in the vicinity of the TOD zone. This site has not been identified, but it would be useful to put both the police station and the fire station into one public safety complex. The second project is to identify the needs of the fire station and police offices and designate an appropriate site for same within a downtown area, providing there is appropriate access. The station will need to be LEED® Gold, meet all TOD zone criteria, and hold all of its stormwater on-site.

5.2.2 Requirements of Proposers

Proposers who are interested in performing the services must exhibit relevant experience with the desired type of work and should be able to emphasize and validate the experience and capability of the specific personnel who will perform the work. So, another hurdle the engineer must face is how to meet the burden of expertise. Firms can rely on developing in-house expertise, can co-venture with other firms to procure various skill sets or include subconsultants who have specific expertise. In all cases, interested parties—corporate or otherwise—and individuals must be fully licensed for the type of work to be performed at the time the response to the RFP is submitted.

The written content of the proposal response document is of paramount importance in getting short-listed for any project. The ability to write a clear, concise, easy-to-understand, well-formatted, and mistake-free response is critical to the success of any RFQ/RFP response. Getting through the proposal process to the interview is often the most important step toward earning the job contract.

To ensure that a uniform review process can be conducted fairly, the proposal packages must be organized as prescribed in the client's solicitation. The outline for an actual RFP solicitation that was used by several public sector clients included the following pages from an RFQ released by the City of Dania Beach in 2017:

A. **Title Page**
   This section includes the project name, RFQ/RFP numbers, due date, consulting firm name, and any other information specifically requested in the RFQ/RFP.

B. **Table of Contents**
   A Table of Contents and a List of Tables and Figures will make it easy for reviewers to locate material.

C. **Letter of Interest**
   This letter should include the name of the person(s) who is authorized to make representations on behalf of the respondent, as well as other pertinent contact information such as job title, address, telephone number, fax number, and email.

D. **Description of the Proposer's Team**
   This section should identify the design professionals, subconsultants, including a project and staffing plan, organizational chart, and relevant resumes.

E. **A Questionnaire or Set of Items within the Proposal**
Each respondent should address the required information solicited by the SIQ/RFQ, which typically includes a management plan, technical plan, information about the team members, and a quality assurance plan. An effective management plan briefly describes any anticipated major challenges (and how they might be addressed, the problem-solving approach, and the communication plan for those doing the proposed work).

It is important to describe the organizational chart and how the work tasks will be completed, including identification of all staff members and their duties. The organizational chart that defines the team should identify those individuals who will be most directly involved with the project and how they will interact with the project and each other (see Figure 5.1 for an example).

F. **Firm Experience**
The team should include recent (defined as within the last five years or so) direct firm experience in the design and construction of capital improvement projects that involved directing a similar or more complex scope of services than those listed in the RFP, under a parallel or higher responsibility level.

G. **Direct Project Experience**
Proposers should submit previous directly related project experience with specificity.

Within the discussion of the people that will be assigned to the job, proposers should identify the anticipated level of participation for each individual in comparison to his/her daily workload activities. Proposers should identify a PM who will be directly responsible for day-to-day communication and coordination with the owner. This person must be able to represent the design team and be tasked with making commitments on behalf of the design team.

The proposer's PM must be duly licensed in the state where the project is located and, if necessary, meet any requirements

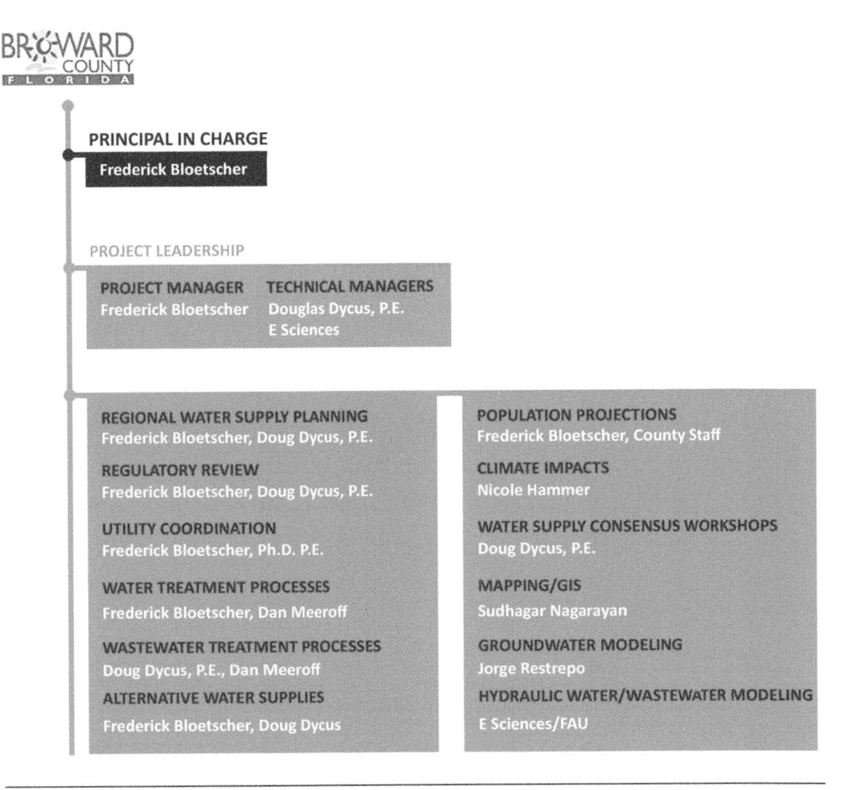

Figure 5.1 Example of an organizational chart

for experience in the RFQ/RFP—for example, 10 years of experience in the management of projects that were equal to or more intricate in technical scope than the services listed in the proposal.

Another consideration will be the description of how subconsultants and subcontractors will be integrated into the organization of the project. It helps to identify experience with an actual previous project and describe how the technical and managerial methodology was applied to keep the project on schedule and on budget.

Identification of key subconsultants who will participate in the project (including reference to the respective tasks they will perform and evidence of their experience and qualifications) is required. The ability to explain how internal and external channels of communication will be employed to achieve successful completion should be demonstrated. Finally, it is critical to make sure that the owner understands the project approach and timeline by providing a (Gantt-style) schedule (see Figure 5.2).

Continuous communication between the client, proposer, and subconsultants is crucial to the overall success of any project, but technical and managerial interaction at all levels is necessary if a project is to move forward smoothly. The proposal response should include how the team is accustomed to working under this type of communication environment. Specific details of previous experiences and applications should be provided, along with a description of how those within the organizational chart will be integrated with the owner's staff. Names of personnel to be assigned (and their duties) should be included along with an indication as to how subconsultants will be able to adapt to this type of working environment.

A summary on how the team plans to ensure proper communication should be included. Details on how any approach to communication concerns has worked well in the past should be documented and previously successful strategies should be briefly referenced.

Included should be a description of the team's applicable quality assurance/quality control program or protocol, which should indicate

Task	Lead	1	2	3	4	5	6	7	8	9	10	11	12
TASK 1 – Project Kickoff, Project Schedule, and Information Collection	PUMPS	■											
TASK 2 – Introduction	PUMPS	■											
TASK 3 – Potable Water Treatment and Source of Supply	PUMPS		■	■	■								
TASK 4 – Wastewater Treatment and Effluent Disposal	PUMPS		■	■	■								
TASK 5 – Operational Goals	PUMPS		■	■	■								
TASK 6 – Describe Existing Potable Water Distribution, Wastewater Collection, and Raw Water Piping Systems	PUMPS			■	■								
TASK 7 – Growth Forecasts	PUMPS		■	■	■								
TASK 8 – Potable Water Modeling	Esciences			■	■	■	■	■	■	■	■		
TASK 9 – Wastewater Modeling	Esciences			■	■	■	■	■	■	■	■		
TASK 10 – Raw Water Modeling	Esciences							■	■	■			
TASK 11 – Recommended Improvements	PUMPS									■	■	■	
TASK 12 – Publish Master Plan	PUMPS												■
TASK 13 – Presentation to Board of County Commissioners	PUMPS												■

Figure 5.2 Typical Gantt-style bar chart schedule

specific steps to be conducted for technical review of any type of deliverable, prior to submission to a client. The team should identify any standard processes used and any success factors that demonstrate and confirm effective outcomes.

Highlighting experience with similar projects is important. A project description should generally be developed in the following fashion (see Figure 5.3 for an example):

- Project location
- A brief project description
- Name of the client
- Date of completion
- Estimated cost and actual cost after completion
- Whether or not the firm was the principal firm in charge of the project, and if not, the name of the principal firm
- Specific service performed by the firm submitting the proposal and the principal in charge (in the case of team submittals or joint venture submittals, identify which individual firm was responsible for the project or portions of the project listed)
- Name, address, and phone number of references familiar with the project and the services performed by the firm
- Names of subconsultants used in the project

Most proposals also require a statement concerning possible litigation or other dispute resolution proceedings (e.g., arbitration or mediation) in which the firm or individual staff is currently involved or has been involved in over the past five years. There should be an emphasis on points of contention as well as results, if available. This information should be summarized with appropriate appendices, including details.

Finally, the proposal should include an approach to providing the scope of services (hence, why a good scope is better than a poor one). This section should demonstrate how well the team understands the issues associated with the projects. The proposal should include a statement as to the project team's understanding, planned project approach, and a tentative time of performance for the scope of services.

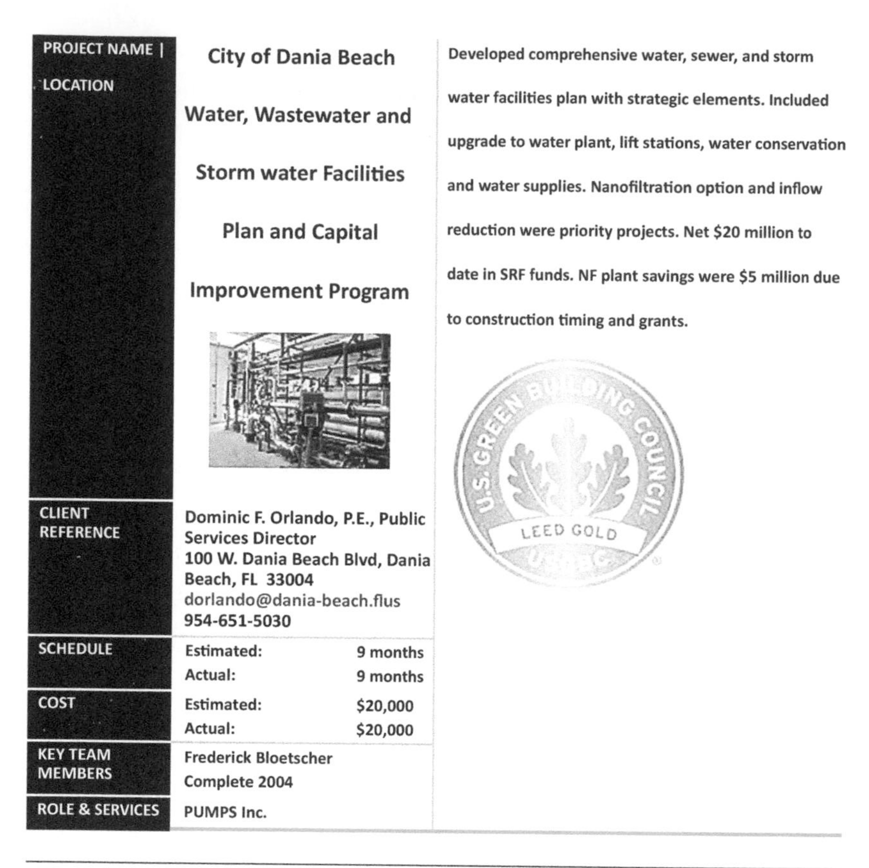

PROJECT NAME | LOCATION	City of Dania Beach Water, Wastewater and Storm water Facilities Plan and Capital Improvement Program		Developed comprehensive water, sewer, and storm water facilities plan with strategic elements. Included upgrade to water plant, lift stations, water conservation and water supplies. Nanofiltration option and inflow reduction were priority projects. Net $20 million to date in SRF funds. NF plant savings were $5 million due to construction timing and grants.
CLIENT REFERENCE	Dominic F. Orlando, P.E., Public Services Director 100 W. Dania Beach Blvd, Dania Beach, FL 33004 dorlando@dania-beach.flus 954-651-5030		
SCHEDULE	Estimated: Actual:	9 months 9 months	
COST	Estimated: Actual:	$20,000 $20,000	
KEY TEAM MEMBERS	Frederick Bloetscher Complete 2004		
ROLE & SERVICES	PUMPS Inc.		

Figure 5.3 Example project description

Issues such as zoning, codes, construction costs, potential permitting issues, environmental issues, standards or code requirements, and any other applicable constraints should be identified, along with proposed solutions since these items might impact the owner's schedule or cost. If the scope is not completely understood, the project may be doomed from the start.

5.2.3 Evaluation of Proposals

The typical selection process leading to identification of the most preferred provider is as follows (based on language from a series of actual proposal requests):

1. Issuance of the request.
2. Period allowed for questions and answers from potential proposers.
3. Receipt of responses.
4. A selection committee will score each submittal in accordance with the rating guidelines and may ask the top-ranked firms to schedule an interview, presentation, or both.
5. The committee will rank qualified firms in order of preference and present its recommendations to the decision makers.
6. The decision makers enter into negotiations with the preferred firm. If negotiations are unsuccessful, then the owner may invite one of the other top-ranked firms to negotiate or reopen the process.
7. Upon the successful completion of negotiations, an agreement will be signed by both parties.

It is typical for the selection procedure to be specified in the RFP, local codes, or statutes—or otherwise detailed in writing for the respondents. The following are common categories used by most agencies for scoring proposals (Dania Beach RFP for Engineering services, 2017):

1. **Company expertise:** Ratings are typically based on information provided with respect to the type of work described within the RFP. Directly related expertise will likely have a higher rating than only limited experience in other types of projects. Level of difficulty and the successful overcoming of strategic challenges in similar projects may also be considered within ratings. Firms with previously successful work, locally or with a well-defined team structure, and functional quality control policies may receive higher ratings.
2. **Previous staff experience:** Ratings are typically based on the experience of the specific staff members who will be involved in the day-to-day design and construction related to the type of work (as described within the RFP). Significant individual experience in performing similar projects may receive higher

points. Limited staff experience will likely result in receiving fewer points.

3. **Current and projected workload:** Ratings typically reflect the workload (both current and projected) at the firm, the number of staff assigned, and the percentage of availability regarding the staff members who are assigned. Current projects that are ongoing in the same offices should be identified.
4. **Office location:** Ratings are based on the ability of the team to execute any level of the contract work and provide subsequent responsiveness in a timely fashion.
5. **Demonstrated prior ability to complete projects on time:** Respondents need to provide a tentative project timetable. This schedule should include both design and construction portions of the project. Respondents will be evaluated on the logic applied to each timeline, interrelationships between project timelines, and predicted impacts to scheduled projects, as well as subsequent proficiency in establishing a streamlined and successful delivery process. Respondents will be evaluated based on previous experience pertaining to successful completions and steadfast conformance to similar project time frames.

   Specific attention will be given to successful strategic and managerial approaches that were utilized to exercise timely project completion, as well as the ability of the respondent to provide full, dedicated attention to each workload priority. Respondents who have demonstrated an inability to complete projects on time will receive a lower rating.
6. **Demonstrated prior ability to complete projects on budget:** Respondents will be evaluated on their capacity to establish competitive and technically responsive projects, as well as the ability to adhere to initial budgets. Comparisons shall be made between initial negotiated task costs and final completion costs. A table that shows initial budget/award and final cost is the best means of presentation. Respondents shall be given the opportunity to explain budgetary overruns, and consideration

> shall be given to scope modifications because of expansions or reductions in original scope. Unjustified budgetary overruns shall reduce point totals.

As noted, price is not permitted to be a consideration in many jurisdictions. Among the reasons price is excluded is to allow local entities to secure the most qualified professional so that any learning curve is minimized.

Most engineers in the private sector also attempt to gain work through submission of proposals that clearly articulate ideas, qualification, experience, and the ability to accomplish the project. A lot of money is spent by engineers to develop responses to SIQs and RFQs, and even more is spent on RFPs that may require full preliminary proposals. Even simple proposals may cost anywhere from $20,000 to $25,000 in staff time to develop—if done carefully. On very large projects, firms may spend a half a million or more.

Many large defense contractors use engineers to create conceptual design packages and convey that data to prospective clients for multi-million-dollar government military projects. The RFP process is similar in all fields, as engineers try to secure work through a proposal.

5.3 NEGOTIATION OF CONTRACTS

Once the project is secured, a contract must be created. Most owners present standardized contracts. If not readily available, engineering organizations can provide these documents. Most engineers are not lawyers and would be advised not to write their own. Such a self-produced document can be construed against one's interest if there are conflicts. Scope, price, and schedule are the critical items that must be defined. Scoping is needed to define the other two factors of price and schedule.

When defining a scope for project contracts, of critical importance is a description of the product and what it is designed to do (which may include product increments or interim deliverables produced during the project). This detailed description can identify missing criteria prior to delivery. Equally critical is the inclusion of specific details of what will *not* be contained in the project. This is part of managing expectations of the client.

The client should be made aware of any assumptions concerning the project or conditions as they exist or are expected to emerge during the project. This includes any time constraints, rules, regulations, or other factors that may limit or slow down the work progress during the project. Once any such requirements are identified and prioritized, the project management team needs to ensure that they are measurable, testable, traceable, complete, consistent, and will meet client expectations. If requirements are not collected in enough detail, it will be hard to set objectives that will ultimately satisfy the stakeholders. At this time, requirements can be included in project baselines so that they can be used to measure progress and compliance.

Negotiation of the scope requires that the engineer develop a rapport with the other party before entering into the negotiations in earnest. At that time, the primary focus should be on the scope of the project and potential issues. Negotiation requires that the engineers establish competence and credibility, which is helped by intently listening to the other party to understand their specific needs and concerns. All negotiations should be concluded on a positive note with respect to scope, price, terms, and schedule. The goal is a win-win/mutually beneficial agreement based on the interests of both parties.

5.4 WORKING AS AN ENGINEER

Once the project is secured, the engineering entity's effort begins. All engineering projects involve teamwork. The best teams are made up of individuals who go out of their way to support each other with an appropriate amount of decorum (respect). An effective team opens multiple lines of communication and frequently brings members together to discuss progress, plan for future deliverables, evaluate concepts, make decisions, rehearse presentations, and vigorously edit written submittals.

Effective team meetings can be critical in keeping the project on schedule. However, if team meetings are ineffective, they can be a source of dissension and erosion of team cohesiveness. A good way to maintain effectiveness within team meetings is to evaluate how effectively the time was utilized. In evaluating the effectiveness of a meeting, the following question should be answered:

- Did it start and end on time?
- Did everyone participate?
- Were important issues discussed and significant decisions made?
- Did you reach a consensual solution?
- Did everyone engage and add to the discussion?
- Did you allow for new ideas to come forward?
- Did you explore these new ideas and include them?
- Were any negative issues brought up?
- Were these negative issues resolved?
- In retrospect, was the meeting efficient and effective?
- Were significant decisions made?
- Were these decisions and related matters recorded?
- Did you stay on the agenda?

No one said it would be easy—and often, just as in real-world situations, conflicts, difficulties, and communication issues arise within teams. Decorum is part of the ethics of engineers. Participants can disagree, but decorum is required.

Teams that discuss issues, share opinions, resolve disagreements, and gain an understanding of all differing points of view are typically those that function well and perform effectively. Therefore, conditions need to exist where every team member is provided an opportunity to speak and be heard. Some team members may be uncomfortable about expressing their opinions, questioning others, or defending points of disagreement in group settings. They will likely distance themselves for fear of confrontation or humiliation. If this happens, one or more important points of view will be silent or lost. So, it makes sense to create some important ground rules for teamwork.

5.5 STAGES IN THE DESIGN AND FABRICATION PROCESS

Once the design professional has been selected by the owner, and all contract documents have been executed, the design begins. There are four stages of the design process:

- **Conceptual design:** Description of what is needed versus what is there
- **Predesign:** Calculation and sizing of the facilities needed
- **Preliminary design:** Plans and specs of the project are to be roughed out
- **Final design:** Bidding documents

The first requirement is to define the needs for the project and, as noted previously, to gain a good scope and make sure that you stay within the scope (avoiding *scope creep*). The design professional needs to clarify the needs and potential options that are acceptable to the owner. This clarification includes getting answers to the following questions:

- What does the client want?
- What is the function of the project?
- What facilities are to be included?
- How many people will be served?
- What codes and regulations must be met?
- What are the environmental considerations that must be met?

The result of these discussions should result in a conceptual design or basis of design (BOD) report. It is important that during the conceptual stage, the operations and maintenance staffs be involved since these are the personnel who will keep the project operational going forward. To operate optimally, the owners need to know what is required. As a result, a BOD report should include:

- Introduction of the problem
- Background information, as well as its needs and current facilities
- Alternatives to be considered in order to resolve the problem
- An evaluation of alternatives
- Recommended possible alternatives

Conceptual design may also include drawings that outline salient features, especially when trying to obtain comments and guidance prior to the design. The owner should provide written acceptance (but not approval) of the conceptual design report and drawings. This acceptance provides the design professional with the approval to move forward with the project. The owner does not want to *approve* the report or plans at this time given that this may cause the owner to assume some degree of responsibility by inferring that some degree of review, calculations, etc., may have been undertaken by stakeholders (which is rarely the case). Once this acceptance has been recognized, the design professional will develop the preliminary design plans and specifications.

Upon receipt of the preliminary plans and specifications, the owner should again provide comments, for the reasons noted previously. The final design plan typically includes information detailing materials, fabrication, and components (from others), as well as a set of drawings. Note that all of these should be directly related to the basis of the design documents that were created initially. After receipt of this acceptance, the design professional will work toward the final plans and specifications.

For a civil engineering project, the final data set may include any or all of following types of documents (as appropriate for the project and in addition to the specifications):

- Cover sheet
- List of drawings and the engineers who made them
- Site plans—existing
- Site plans—proposed

- Site plans—changes highlighted
- Water and sewer plans for site
- Stormwater plans
- Stormwater disposal details
- Parking plans
- Landscaping plans
- Floor plans
- Elevation plans
- Roof plans
- Structural plans
- Structural details
- Plumbing plans
- Lighting plans
- Electrical plans
- HVAC plans
- Details

5.5.1 Documentation and Reports

Documentation is a key concern for engineers—both for protection and to demonstrate ethics. Being sloppy in documentation raises legal and ethical issues. Writing technical reports requires the same skills and attention to detail as has been discussed throughout this chapter. Many organizations will also adhere to a set of strict guidelines that govern the format of these kinds of reports. The final technical report for a project will typically have an outline such as the following:

- **The cover page**—should contain all of the essential elements including the title of the project, the contributors' names, the date of submittal, etc.
- **A table of contents**—provides the structural outline of the report so that finding the exact location of key items is made easier for the reader.
- **A list of figures and tables**—provides the same thing for the *graphics* included in the report. If the captions were made

appropriately, then these are simply lists of captions with the page number locations. The professional engineers (P.E.s) who designed certain aspects will sign, seal, and date these visual elements as to their veracity.

- **The introduction**—to the report describes the project goals, the location of the project, and the objectives. It must start strong with scope and objectives clearly presented. This is where the engineering team will note assumptions and documentation that is relied on in the design. The introduction should fully express the primary purpose and scope in its context at the beginning of the report. The introduction should also present a clear statement that demonstrates how the report will track the fundamental, secondary, and implied problems, questions, and issues described within.
- **The body of the report**—contains all the supporting information to address the major items subdivided in sections that are all related to the goals of the project. These typically include: existing conditions, discussion of alternatives, selection of alternatives, recommendations, issues that need owner input, and risks as necessary, among others. The text within the body of the report should convey a professional level of knowledge of the subject matter with no relevant content omitted and no incorrect material presented. The report's focus must be clear to the reader such that the paragraphs are logically and coherently placed upon each other through the complete and fluent use of transitions. The report must finish with a logical conclusion that is supported by the data that has been presented or referenced.

The writing should exhibit substantial depth and complexity of thought, but also should be supported by well-developed ideas, analysis, or evidence. All content within the narrative should tie back to the original purpose and goals of the project. Facts should be presented in a logical sequence, and sections must effectively transition between topics and different authors. Some other characteristics of a well-written body are as follows:

- Seamlessly incorporates and explains the accuracy and relevance of data, evidence, quotations, paraphrases, and visuals
- Offers evidence from a variety of sources, including counterarguments, contrary evidence, and quantitative analysis
- Presents data in graphical, tabular, or sketch format—following all rules for graphical format, including proper units and labels
- Spelling and grammar checked
- Sentences consistently communicate thoughts clearly, while relatively free of sentence level patterns of error by using a technically sound sentence structure that is varied, convincing, nuanced, and eloquent with an appropriate tone
- Evidence of editing

Most important, the supporting material must build toward an effective conclusion in order to finish strong with a reasonable summary and/or recommendations presented, as justified from the body of the report being used. A complete reference section should be included that cites and formats literature sources accurately and consistently.

Each report should be designed to serve as a stand-alone technical memorandum (TM) or contain supporting materials that are designed to be a complete summary of the design for a particular aspect of the project. It is expected that the reader will be able to take this information and replicate the work. Hence, all calculations, codes, and drawings must be included. A written description of methods, assumptions, and application of technical data is required.

5.5.2 Failure to Write TMs and BOD Reports

A couple of examples will illustrate why BOD reports are so important to the successful completion of engineering projects. These two examples involve water treatment plant expansion projects, but the issues are generally related to electrical and structural components. In one example, the design report helped the engineer because it was complete and well-crafted. In the other example, the design report was found lacking. However, the public interest was not served in either case. There was an opportunity to rectify the situation.

CASE STUDY 5.1

Membrane Treatment Plant

The first case study concerns the design and construction of a large membrane water treatment plant for a utility. The consultant chosen for the project had designed prior plans for a similar construction, including one in a neighboring municipality. The project cost was expected to be over $20 million and involved a short construction time. The construction consulting fees were approximately $1.2 million.

The services to be provided by the consultant included a design of the facilities, contract administration, daily inspection, services during bidding, negotiation of change orders, and review of the schedule. The utility's engineering staff had some experience with construction projects and was providing periodic inspection and project coordination on the less complex components of the project. As a result, the consultant was receiving some direction from the utility staff of the project on a routine, consistent basis, including reviews of the drawings and discussions with the operational staff.

The project was constructed in two phases. The first phase was nanofiltration, with the reverse osmosis phase to follow in a time frame of three to five years later. The design incorporated all of the piping and electrical work that would be necessary to satisfy the expansion phase (reverse osmosis), which was similar, but involved different water quality.

The first phase of the construction was difficult. The contractor was litigation-oriented, and the construction timeline was not realistic. As a result, the contracted construction time became unachievable and significant animosity developed due to the delays. The local government ignored the recommendations of its project management staff to terminate the contractor and instead decided that more personnel should get involved. After much effort, the project came in more than a year late, and litigation with the contractor ensued. The animosity that was

continued

created between the utility and the consultant during the first phase was such that the design consultant was not selected to complete the second phase, despite obvious familiarity with the project.

A second consultant was hired to perform the second phase and design the upgrade. The new consultant based the design on a project that had been constructed elsewhere. However, it was revealed later that little design work had occurred, since many of the parameters used in this project were exactly the same as the previous project, particularly with respect to the mechanical and electrical system design. The original design called for the pressures on the reverse osmosis pumps to be 405 psi. The electrical system was not capable of handling this pressure, so a claim was made by the utility against the initial design consultant. The claim was for $1.5 million, which exceeded the amount paid to the initial consultant to design the plant.

However, upon review of the claim by the client's lawyers and an external consulting engineer, it was determined that the new consultant had not followed the design parameters of the initial consultant. The original design was achievable and therefore there was no reason to deviate. Worse, the second consultant knew all of this—having been provided with a copy of the BOD report that was developed by the initial consultant—but had not incorporated it into the second phase design. Further, the design that was submitted was *cookie-cutter* identical to a design for another facility.

The second consultant provided no BOD report. Upon revelation that the initial design concepts had neither been followed nor evaluated, and that a design copied from another plant was submitted without consideration of the initial design, the legal team terminated the second engineer. Therefore, a claim was made for back fees. The first engineer had the $1.5 million claim dropped because of the BOD report.

EXAMPLE 5.3

Failure to Document

What can possibly go wrong if you and your design personnel are not properly licensed, and you lie about it? Both are significant ethical violations that are punishable by state engineering licensing boards. This example involves the design and construction of a large wastewater treatment plant by a large, multistate consultant who had been working in the state for more than five years. The consulting services for the project included design and construction observation services.

The construction involved a poured-in-place, reinforced concrete tank. It was noted that the contractor constructed the tank in an odd, checkerboard fashion, but because the staff had little experience with large tank construction, little notice was paid. Upon completion of the tank, hydrostatic testing commenced. During the hydrostatic testing, it was found that consistent, diagonal cracks were visible at ten-foot intervals along the perimeter wall. The direction of the cracking diverted midpoint of the wall. Water from the tanks was visibly leaking from the walls.

The consultant was tasked to investigate the problem and provide a response. His response was that "the cracks will heal" and "it is the contractor's responsibility." The consulting engineer declined any further responsibility for this work.

The second consultant determined that the cracking was caused by insufficient longitudinal steel being placed in the wall—a structural design problem. Two additional consultants were asked to review the project; both reached similar conclusions. As a part of the second evaluation, it was found that the contractor was concerned about the amount of steel used and had developed the odd construction sequence to minimize shrinkage problems.

The design consultant ultimately denied that the remediation was necessary and sued when the corrections that were recommended by the second consultant were made by using the funds that were due to the original design consultant but had not yet been paid.

continued

As litigation ensued, it was found that the original design consultant had never developed a BOD report, nor were available calculations provided. The only calculations that were initially provided were references to an American Concrete Institute (ACI) standard (304) which was applicable to parking garages (at the time). A tank design should have referenced a different ACI standard. This use of an irrelevant standard was revealed during depositions. Worse yet, it was determined that the lead structural design engineer did not have a P.E. license, nor did the consultant that reviewed the design. Ultimately, the lack of licensure and the use of inadequate design standards was determined to be an ongoing issue that affected other sites as well. Once other clients determined they had experienced similar issues, the consultant was soon out of business.

5.6 SUMMARY

The goal of this chapter is to reveal how engineers can approach the challenge of obtaining work. As engineers gain greater responsibilities over time within their firm, they can be expected to become *rainmakers* (those who can secure contracts and revenues to keep the staff employed). The process is similar within all of the public sector and, at times, the private sector appears to be adopting some of these practices. Scoping the project and working together are major responsibilities, regardless of the type of entity you may work for.

REFERENCES

Bloetscher, Frederick. (1999). Looking for Quality on an Engineering Consultant. *American City and County*. December, 1999. p. 36.

———. (1999a). What You Should Expect from Your Consulting Professionals (and How to Evaluate Them to Get It). *Water Engineering and Management*. October, 1999. pp. 24–27.

———. (2011). *Utility Management for Water and Wastewater Operators*. AWWA. Denver, CO.

Bloetscher, F. and D.M. Meeroff. (2015). *Practical Concepts for Capstone Design Engineering.* J. Ross Publishing. Plantation, FL.

Mantell, M.L. (1964). *Ethics and Professionalism in Engineering.* Collier-MacMillan Ltd. London, UK.

Popkin, R.H. and A. Stroll. (1993). *Philosophy Made Simple.* Broadway Books. New York, NY.

PROBLEMS

1. Find several RFPs that were submitted in your field. Compare what was asked for in each. Put this in a table. Compare and contrast each RFP.
2. How are engineers and fabricator solicitations different? Compare how these differences vary in public and private entities.

This book has free material available for download from the Web Added Value™ resource center at *www.jrosspub.com*

CHAPTER 6

ORGANIZING FOR WORK

Having reviewed the history of the profession and how the canons, codes, creeds, and other guidance documents were developed in order to help engineers protect the public, a question might arise as to the different types of organizations that exist, especially since many practitioners may want to create their own firm at some point in time. This chapter presents those options.

LEARNING OBJECTIVES

- Define the types of organizations that exist
- Outline the benefits and risks that are associated with each type of organization

6.1 TYPES OF ORGANIZATIONS

There are four major ways by which services are offered: individuals, sole proprietorships, partnerships, and corporations. Each has its advantages and disadvantages. Note that individuals who work alone have the greatest number of barriers and the highest risk factor. That should be a consideration for all engineers.

Individuals who offer services are rarely utilized for many reasons. First, while one person is responsible for everything, it can become a problem since there is no other person to share accountability. Since personnel resources are limited, most large entities resist contracting with an individual. Risk is high, control is high, and while the responsibilities are known, the initial capital acquisition is extremely limited and the success of the entity lives or dies with the individual. It is also a major challenge to secure professional liability insurance in an instance such as this. Hence, for the client, the risk is high when hiring individuals to do major work.

One way to get around the obstacles of hiring a single individual is to create a public entity called a *sole proprietorship*. This type of entity is simplified, and all liabilities are the responsibility of the owner. A sole proprietorship can have multiple employees (preferable to involving only one person), and from a tax perspective, the profits and losses can be advantageously attributed to one party. However, individuals and sole proprietorships are often not viewed as being significantly different by most large organizations, even though the sole proprietorship might have employees (possibly many employees). The same concerns with capital acquisition and professional liability insurance exist in both instances.

A second legal entity is a *partnership*. A partnership is defined as a legal relationship between two or more people who enter into an agreement to work together. This type of entity is common among law firms and engineers. In this type of entity, the risks and rewards are shared, as is control of the entity. Initial capital acquisition is less limited and interest can be purchased by other partners at any time. Liability insurance is usually not difficult to acquire.

Vance v. Ingram is a court decision in which the terms of the agreement are rigidly defined. Each partner can enter into a contract and all partners inherit the risk. Benefits to the parties within the entity lives on until dissolved. Thus, if a partner dies or wants out, those members among the partnership would have first right to acquire any shares so affected.

A *Limited Liability Corporation* is a legal entity much like a partnership where the partners contribute money and talent, share in profits and losses, and share in tax obligations. In this case, the partnership is limited in that the total losses cannot exceed investments made by the partners. However, the limited partners have no vote or say in daily operations. The LLC can be dissolved for any number of causes, including:

- Insanity of a partner
- Death of a partner
- Principles addressed in documents
- Bankruptcy
- A specified date
- Mutual agreement
- Court orders
- Expulsion of a partner
- Interests acquired by other partners

In dissolving the LLC, creditors must be paid from assets or loans. In addition, advances must be paid and capital investment must be returned before distribution of any remaining assets.

More common are the true corporations—S or C Corporations issue stock, which entitles each shareholder to a vote on the Board. The stock is 100% transferable—no permission needed—and shareholders may use proxies for votes.

Corporations can easily raise capital (by selling more shares of stock) and can secure and convey real estate. The shareholder's risk is legally limited to the value of the stock investment (only as corporate value is based on the value of the stock, which is based on actual or anticipated profits). The corporation takes on all risk and liability while shareholders gain rewards (or in some cases, monetary losses). Anyone can buy and/or sell shares and the entity lives on until it is dissolved.

There are two corporate structures that are based on taxes. *S corporations* make up over 90% of all corporate entities. By definition, S corporations pay no income taxes since these tax obligations are transferred to the shareholders. The reason that they are not taxed is because the shares are rarely transferred and the tax obligation is easily transferred to the shareholders as a result. Most partnerships, LLCs, and S corporations routinely reduce their year-end *profit* by distributing bonuses and deliberately making year-end purchases to reduce income tax obligations of the shareholders.

This is why any discussion of corporate income taxes by politicians immediately exempts the majority of corporations because only *C corporations* are subject to income taxes. An example of a C corporation is the Ford Motor Company, which has issued millions of shares of stock that are actively traded. Hence, the tax obligations are kept within the corporation since they are not easily transferred to any individual because millions of shares are bought and sold each day. There is no way to tell who made the profits and when.

There are many reasons that an engineer would create an S corporation. One is a distinct, court-recognized existence, which helps protect individuals from personal liability that can result in the loss of personal assets during litigation. As can be seen, S corporations are preferred when businesses provide a service to the public (i.e., consultants). The S corporation does not have significant start-up costs and would not need to make major equipment purchases before beginning operations. Thus, S corporations can raise sizable capital without a great deal of effort and expense.

Note, however, that S corporation entities are not suitable as an estate planning vehicle since control is ultimately in the hands of the stockholders. In a planned gifting scenario, once the majority control passes to children from their parents, children can take full control of the company. If tax status is compromised by either nonresident stockholders or stock being placed in a corporate entity name, the IRS will revoke status, charge back taxes for three years, and impose a further five-year waiting period to regain tax status.

Joint ventures are commonly found among organizations that are seeking to work together on a project. A joint venture involves two or more firms that enter into an agreement, which is generally project-limited. As a result, the agreement dissolves upon closure of the project. Engineers may choose to enter a joint venture in order to be considered for participation in larger projects; to offer specific experience; or if the original firm lacks familiarity with local customs or regulations, to fill a void in expertise or skills (localized surveying, for example). A design/build joint venture between contractors and engineers to construct a project would be a common example.

Table 6.1 outlines the general types of organizations as well as their advantages and disadvantages.

Table 6.1 Advantages and disadvantages of various types of organizations

Type of Entity	Limited Liability Protection?	Tax Treatment	Level of Government Requirements
Sole proprietorship	No	Taxed at personal tax rate	Low
General partnership	No	Taxed at personal tax rate	Low
Limited partnership	No; Limited partners only	General partners taxed at personal tax rate	Medium
S corporation	Yes	Taxed at personal tax rate	High
C corporation	Yes	Must pay corporate taxes and beware of double taxation of dividends	High
Limited liabiity corporation	Yes	Can choose how you want to be taxed	Medium

6.2 ONGOING OPERATIONS OF A FIRM

Once an engineer (or group of engineers) creates an entity, there are a number of ongoing activities that must occur. The most obvious one is acquiring work which was outlined in Chapter 5. Assuming the entity acquires work, there are two additional issues to address. The most important is getting paid. That seems simple, but the reality is that most engineering services receive payment *post completion* or when the work is partially completed. As a result, an entity is always getting paid after the work is done and expenses have incurred. To receive pay, an invoice needs to be created and reviewed by the client, then a check or bank transfer is authorized. Many governments have 30 to 60 days to pay once an acceptable invoice is approved. Therefore, most engineers will find that payment occurs 60 to 90 days after the work is complete and all bills are paid. Those payments typically necessitate the use of reserve funds or short-term loans.

Some might argue that bills will be met after the engineer gets paid. Ask your employees about that; it is not acceptable. Payroll and payroll taxes are among the most important expenses each month. Payroll taxes (Social Security, Medicare, and withholding) are required to be submitted to the IRS 10 days after the month is complete—and yes, the IRS watches this. Do not pay early! That is an even bigger issue because, believe it or not, the IRS assumes that no one pays early. State and local governments may also have payroll, unemployment, and a variety of other taxes that must be paid periodically. Like the IRS, they expect prompt payment. The IRS and certain states require monthly, quarterly, and annual filings as well. It is critical that these filings match up with deposits.

Insurance premiums must be paid before the insurance coverage is issued. Professional liability insurance is critical and mandatory, but so is general liability, accident coverage, and other forms of indemnity. Many entities require unemployment and workmen's compensation insurance verifications. Insurance must be applied for, and the cost will vary with the type of clients and total revenues received. Greater revenues and a greater number of private clients will lead to greater costs.

To address issues concerning the IRS, accountants are normally hired to prepare corporate taxes. Discussion with an accountant to create a year-end plan is important. Most S corporations pay no income taxes because the end-of-the-year balance goes via Schedule K to the shareholders' personal taxes. Most accountants try to zero out (or make negative) the end of the year statement since personal tax rates are higher than those for C corporations and investments.

Other expenses come at a lower priority. These include equipment (such as computers, which the accountant can depreciate), office leases, furniture, vehicles, etc. These are necessary to start a business and must be considered. They also must be included in the overall pricing of jobs. Note that certain support personnel such as home office staff, accounting employees, lawyers, and officers who do not work on projects are considered *overhead*, so these costs must be divided up among projects as a multiplier on hourly rates. Most consultants find that the multiplier is around three—the actual cost is three times the cost of the wages of the engineers working on the project.

6.3 SUMMARY

The goal of this chapter is to outline the type of organizations that engineers can create to perform work. Note that most states require engineering firms to designate an engineer to be the responsible party who is in charge of the entity. Therefore, a firm cannot exist without a licensed professional.

PROBLEMS

1. Obtain information on five engineering companies in your field. What types of organizations are they? What states are they incorporated in?
2. Go to a site such as mycorporation.com. What are the steps needed to create your own company?

This book has free material available for download from the Web Added Value™ resource center at *www.jrosspub.com*

CHAPTER 7

EXAMPLES OF ETHICAL CONUNDRUMS FOR ENGINEERS

Now that we have reviewed the history of the profession and the canons, codes, creeds, and other guidance documents that were created to help engineers make good decisions in order to protect the public, the next question is: what could occur to complicate matters? The goal of this chapter is to provoke thought—engineers can do great things, but too often they are self-constrained when looking at the larger view of the impact of their efforts upon others.

LEARNING OBJECTIVES

- Define and understand the impact of misconduct upon licensure
- Recognize ethical questions that arise in a larger arena (such as working with elected governmental officials)
- Identify and address any perceived impact of projects on others
- Embrace challenges that engineering practitioners face as agents of a social change society

7.1 THE CHALLENGES OF THE ENGINEER'S CREED

As outlined in Chapter 2, the National Society of Professional Engineers (NSPE) publishes the Engineer's Creed and Code of Ethics. The creed outlines the need for duty (per Kant, 1785) and public welfare as priorities. Through their actions, engineers are agents of social change in the world—a responsibility that should not be taken lightly. The American Society of Civil Engineers (ASCE) and NSPE now include the concept of sustainability within their codes of ethics, which reinforces the goal of public welfare.

7.2 AGENTS OF SOCIAL CHANGE AND SOCIAL RESPONSIBILITY

Engineers are agents of social change. If there is any question about that, consider how society would have developed without clean water, sewerage, transportation, or energy systems. Then add to that the growth of civilization with advancements in communications, computers, and numerous machines; it becomes obvious that engineers create things that continually impact the human dynamic.

Even within a singular discipline this can happen—Henry Ford's introduction of the industrial engineering concept of the assembly line opened the door for a new, cheaper, and cleaner means of transportation. No longer did cities have to suffer under the auspices of endless horse manure that created unhealthy conditions, especially in the summer months. Every development that comes along creates a new change thereafter. Computers are another example. The way the United States was able to improve efficiency in the 1990s was based on computerization and the use of robots to replace slower humans and less efficient workers in factories.

These changes, however, also create disbenefits. In the former example, thousands (in New York alone, 15,000) of sanitation workers lost their jobs—shoveling manure—when cars came along. Likewise, the Concord coach industry disappeared when Henry Ford developed the assembly line to make cars, cheaper, faster, and in greater numbers.

While cars were a major advance, no one paid much attention to those workers who were displaced. With robots, thousands of automobile workers lost their jobs. The use of robots was the largest impact on the loss of jobs in the manufacturing sector in the United States, not the expansion of production in China (Brown, 2020).

We, as a nation, are still suffering from the mass displacement of manufacturing workers that is attributable to robotics. As a result, much of the current political discourse has underpinnings associated with the loss. The same can be said for the development of fracking technology, which provides less costly fuels and therefore is generally perceived as preferable to coal. We have not sufficiently addressed the plight of displaced coal miners, but we all agree that gas is cleaner and more efficient.

7.2.1 Social Responsibility

It may seem that engineers must be holistic in their analysis of any impacts of the tools they create. This is called *social responsibility*, which is one of the connectors between ethics and engineering. The social responsibility aspect is relevant, as noted previously, in evaluating the impact that engineered products and processes have on society. This does not necessarily mean that these advances are bad, but as citizens and practitioners, we are obligated to recognize what the change means to society. Social responsibility matches to the concept that the Engineers' Council for Professional Development, now ABET, adopted in 1974 that holds the paramount obligation of engineers is to the health, welfare, and safety of the public (https://www.encyclopedia.com/science/encyclopedias-almanacs-transcripts-and-maps/engineering-ethics-overview).

Ultimately, the goal of any engineer is to design and construct products that benefit society (with someone else likely marketing them). However, dedication to social responsibility requires taking into consideration societal needs. In this way, social responsibility must involve professional responsibility. As a result of this nexus between professional and social responsibility, the NSPE and other societies have been able to develop their own codes of ethics. The NSPE has refined the responsibilities of engineers down to:

- Hold paramount the safety, health, and welfare of the public
- Perform services only in their own areas of competence
- Issue public statements only in an objective and truthful manner
- Act for each employer or client as faithful agents or trustees
- Avoid deceptive acts
- Conduct themselves honorably, responsibly, ethically, and lawfully, so as to enhance the honor, reputation, and usefulness of the profession

Codes of ethics are typically presented as a set of guidelines to follow, but the literal and actual responsibilities place duties and obligations on one individually and as engineering practitioners. However, while the codes can guide us, exactly what to do in a given situation is not always obvious because in each case, difficult ethical situations require moral reasoning and conflict resolution.

The ethical dilemmas that are being faced by professional engineers (P.E.s) are more difficult to resolve than generally understood, given that the concept of *doing the right thing* and protecting the public health, safety, and welfare often fall into an area that is ambiguous (Marcy and Rathburn, 2015). Traditional ethics systems such as utilitarianism, deontology, and virtue ethics (as discussed in Chapter 2) may provide an opportunity to construct arguments for or against certain policies or courses of action—and to determine whether an argument has been resolved satisfactorily.

In contrast, a *bottom-up* approach emphasizes the need for careful analysis of the particulars of a given situation. This restricts the use of broad moral principles (Jonsen and Toulmin, 1988). However, intuition is not a reliable method for making ethical decisions (Barker, 2008), so codes of ethics were developed to provide a framework for making ethical decisions.

Marcy and Rathburn (2015) note, however, that codes of ethics often tend to be *backward looking* and do not account for the social change that engineers typically create. Rapid advances in technology often create unanticipated ethical dilemmas that may include unintended consequences, while providing positive impacts on society. Significant disbenefits may also be introduced (Marcy and Rathburn, 2015). Herkert (2001) notes that today, the issues in engineering ethics range from micro-level questions about the everyday practice of individual engineers to macro-level questions about the effects of technology on society.

Instead, they suggest that ethical considerations must be included in every step of the design, documentation, and deployment process to help anticipate and mitigate negative consequences. One approach is to conduct a *social impact analysis (SIA)* as a formal part of the engineering design documentation process. SIA is a forward-looking methodology that analyzes the potential ethical consequences of a design. A general outline of the steps that are required to develop an SIA is as follows (Marcy and Rathburn 2015):

1. "What need is it intended to fill?
2. Who are the parties responsible for creating and deploying the design, product, or concept (DPC)?
3. Who will be held responsible if the DPC fails?
4. Who are the stakeholders, both direct and indirect?
5. Conduct a cost-benefit-risk analysis
    a. What are the risks?
    b. What are the costs?
    c. What are the benefits?
    d. What is the impact on the environment?
6. What are the potential ethical consequences to all stakeholders, both positive and negative?
    a. What can be done to mitigate or eliminate negative consequences?
    b. What can be done to maximize positive consequences?
7. Provide a critical discussion for each potential ethical consequence.

8. Identify the decisions that must be made to justify the deployment of the DPC?
    a. What can be done to ethically minimize risks to the stakeholders?
    b. What can be done to ethically minimize costs to the stakeholders?
    c. What can be done to ethically maximize the benefits to the stakeholders?
9. What is the right thing to do regarding each decision?"

As noted by Nguyen (2013), there are two main reasons why engineers often stray from a given code of ethics: overconfidence and impatience. These connect well with the main factors in any failure noted by Abkowitz (2008):

- Communication failure is involved in every disaster
- Financial aspects contribute to most human-caused disasters
- Shortcuts in regular procedures contributed to a disaster
- Design and construction errors are a key factor in accidents
- It is impossible for every engineered project to be designed and constructed perfectly

Conflicts of interest occur whenever engineers represent multiple internal and/or external stakeholders. At each stage of any career, loyalties may appear to change. Those in the private sector have a fiduciary responsibility toward profitability of the firm.

However, firm profitability cannot come at the expense of the health, welfare, and safety of the public (Marcy and Rathburn, 2015). Tattershall (n.d.) notes that there are times when the engineer is placed in a position of having to choose between career/monetary success and the good of the community, the customer, or the public as a whole. However, there is no acknowledgment of the lack of protection of engineers who are asked to put their careers and livelihoods on the line to do the right thing.

Safeguards must be put in place to ensure that engineers are protected under such circumstances. Otherwise, few will be expected to step forward, which puts all at risk. On the other hand, the public assumes that engineers will design in order to mitigate risk, thereby creating the equitable distribution of minimal risks and significant benefits (Shrader-Frechette, 1985, 1991).

Professional organizations are improving at providing practicing engineers with the type of continuing education that is needed to make sound ethical decisions. Engineers might assume that these guidelines should be used to provide moral reasoning as well as a compelling justification for action. To this end, there are several ethical conflicts that engineers often encounter including:

- Safety
- Compliance with regulatory requirements
- Confidentiality of private clients
- Environmental impacts, noting environmental does not specifically mean ecosystem, but includes the built environment
- Data integrity
- Conflict of interest
- Societal impact
- Uncertainty and/or risk

In addition, there is a fiduciary expectation in that engineers will keep promises with respect to performance of contracts and other duties—i.e., honesty and integrity. Implicit within this is a duty to compensate for the injuries that one has done to others; to be grateful for benefits that have been provided; to foster goodwill, wisdom, health, and security; as well as the duty not to harm others physically or psychologically. The concept is a duty to not take a free ride on society, either professionally or personally. More specifically, this involves taking only the appropriate benefits from the efforts that have been undertaken. When applying moral reasoning with prima facie duties, there are two kinds of outcomes:

- Cases where duties *do not* conflict
- Cases where duties *do* conflict

Among the questions one might pose is what is the best way to determine what outcomes are desirable for a given situation. This represents a utilitarian view that suggests identifiable obligations—basically what to do or not to do. As a result, when evaluating the ethical issues, there is a priority of certain outcomes, such as:

- Public safety normally overrides all other prima facie duties. The goal is to avoid creating harm to one person to save another.
- Fidelity overrides beneficence.
- Beneficence and nonmaleficence in relation to lasting positive outcomes override prima facie duties that provide short-term pleasure or pain.

But where does that leave us with concepts like autonomous vehicles that could, and predictably will, eventually injure someone due to happenstance at some point in time?

7.3 MORE COMPLICATED ISSUES

In Chapter 3, the focus was on relatively straightforward issues associated with misconduct—violations of written rules and statutes. There are, however, far more complicated challenges that engineers could face. Some of these are surprisingly easy to analyze, yet the outcomes may be difficult to accept. This is one of the conundrums that engineers must constantly face—sometimes doing the right thing may not be the easiest or most desirable thing to do. So, looking at a series of example case studies might help engineers identify pertinent issues while navigating these challenges.

7.3.1 Fixed Budget Challenge

Suppose you are a registered P.E. and you are asked by your client to design a bridge within a budget of $50 million. After doing a study you determine the following:

1. The ideal bridge can be built for $70 million

2. There are no options that you can find that will meet the required codes and still cost only $50 million

After learning of these results, the client says, "If you can't design a bridge that we can build for our budget, we will fire your firm," which may cause eventual layoffs within your organization, possibly including you. What do you do? The code of ethics for engineers requires that the safety of society is of *paramount importance*. The conflict exists between duty to society and loyalty to one's own career and the welfare of other fellow employees, but all codes and laws dictate that the welfare of society comes first. At this point, the client must be educated as to why the optimal cost is $20 million higher than the proposed budget.

Often these types of budget projections are based on outdated data and do not account for inflation. If these expectations cannot be adjusted, there is no choice—morally, the practitioner must withdraw from this assignment. This is a perfect example of being faced with unrealistic expectations from a client.

7.3.2 Changing the Design

Suppose a registered P.E. is asked by a client to design a long, large diameter pipeline. After extensive study and evaluation—including soils, vibration, and corrosion studies, along with discussions with various other professionals and contractors—it is determined that the only appropriate pipe for the job is ductile iron pipe. Ductile iron is expensive, but given what has been learned, this recommendation is presented to the client.

A supervisor (who is also a P.E.) reviews these plans and changes the piping to pre-stressed concrete (he is friends with the local pre-stressed concrete pipe manufacturer, but he does not note that they have previously discussed the issue). He wants to make the change to the design and then expects the original design engineer to sign and seal the plans containing his suggested changes. Should the engineer refuse, realizing that refusal may mean he or she will be fired?

The code of ethics for engineers requires that the safety of society is of paramount importance. The *conflict* exists between the duty to society and to the firm/supervisor. Within all codes and laws, the welfare of society comes first. At this point, a discussion must occur where the engineer can present his or her case. Perhaps there is new information that may cause an adjustment to the original decision. Engineers should be receptive to new information that may change their decision. If not, the client should be notified of the issue, and in some cases, if the change to the design puts the public health at risk, the licensing board should be notified.

7.3.3 Product Liability

An engineer approaches the boss with information that the team is responsible for failure within a device. The boss says, "Just replace it with a fixed design. There is no need to tell the client about the change because it could hurt the relationship with the company." What should occur?

There are many issues here and so much actual history on how this can go wrong. Let us start with *informing the client*. NSPE Canon 5 specifically addresses this issue, as do many state laws, within the statement "avoid deceptive acts." Because the boss is not telling the client and is telling the engineer not to, the boss is in clear *violation* of *avoiding deceptive acts*. Likewise, so is the engineer if he or she ignores the issue because that puts the engineer in violation of Canon 5 as well. This occurs because deception embodies two modes.

Deception by *commission* occurs when a person deliberately tells a lie, such as when a person reports data that is known to be false. Deception by *omission* occurs when one omits something that another party has a right and interest in knowing. Either way, both are violations. The supervisor wanted the engineer to *omit* something because doing so will help the company. Likewise, omission is a clear violation of Canon 4—"Issue public statements only in an objective and truthful manner."

Canon 1 says to "hold paramount the safety, health, and welfare of the public." The question is whether the failure of the device could create a public health, safety, and welfare issue. If so, the failure to notify the client is a major violation of this canon and statutes rule.

Canon 4 directs engineers to "act for each employer or client as faithful agents or trustees." How does failing to notify the client meet this canon? It does not; thus, it is also another clear violation. Remember that failing to conduct oneself "honorably, responsibly, ethically, and lawfully, so as to enhance the honor, reputation, and usefulness of the profession" is noted in Canon 6.

Ultimately, the correct answer is to notify the client of the issue, as well as what correction has to be made along with why and how this decision mitigates the condition. These actions show responsibility, integrity, and courage. It should also result in goodwill between company and client.

Product safety is an issue, and the fact that engineers who are not licensed and can be found performing under the auspices of a corporation does not excuse this behavior. A couple of examples involving a product liability issue are worth noting.

EXAMPLE 7.1

General Motors 2005–2014

In 2014, GM issued a recall for automobiles because of a defect noted in the power steering column. The National Highway Transportation Safety Administration indicated that complaints had been filed regarding deaths associated with this problem (Bomey, 2014). A Denver auto accident attorney pointed out that in 2004, a full decade before GM began its massive recall of cars with the defective ignition switch, GM knew about the problem and the potential deadly consequences.

While GM did not disclose whether the power steering caused any of these issues, the day after the recall was issued, GM's new CEO Mary Barra was expected to testify before Congress concerning the company's failure to fix the faulty ignition switches.

continued

Reuters, a news organization, obtained what appeared to be a document containing a series of 2005 emails between GM engineers in which they were debating whether to make a design change to the ignition switch (Lienert and Thompson, 2014). The change would have cost an extra 90 cents per unit as well as additional tooling costs of $400,000 (Lienert and Thompson, 2014).

GM decided against the recall and retooling effort, with internal documents involving a cost-benefit analysis that showed it would cost less money to pay for the damages in lawsuits than it would to recall and fix the millions of cars with the faulty switch. In testimony, Barra apologized to all GM consumers and in particular the families of victims while promising to change GM's culture from one focused on cost to one focused on consumer safety. In a related note, Ford had a similar issue in the 1980s with a switch in their steering column. Ford recalled and repaired those cars. GM did not, an action that reminded the public of the Ford Pinto debacle 20 years earlier.

EXAMPLE 7.2

Ford Pinto

During the gas crisis of the 1970s, there was a move by American automakers to manufacture smaller compact cars that would save gas and better compete with those offered by Japanese automobile manufacturers. Every American automaker made at least one such vehicle. Ford's offering was the Pinto. In 1973, reports of Ford Pintos being consumed by fire during low-speed rear-end collisions were received by Ford's recall coordinator's office (Giola, 1992).

A series of lawsuits emerged. In the *Grimshaw v. Ford Motor Co.* case, a 1973 internal memo was exposed from Ford's Environmental and Safety Engineering Division that appeared to be a cost-benefit analysis entitled *Fatalities Associated with Crash Induced Fuel Leakage and Fires*. It was eventually prepared for submission to the National Highway Traffic Safety Administration in support of Ford's objection to proposed stronger fuel system regulation (Grush and Saundy, 1973).

continued

Introduced into evidence, this damning memo implied that Ford was "trading lives for profits." The jury agreed and ruled in favor of the plaintiffs, awarding an amount of more than $125 million. It was publicly revealing how corporate America can operate and how egregiously certain decisions can be made.

Remember, corporate America has a fiduciary responsibility to protect the investments of their stockholders by maximizing profits, but not at the cost of failing to protect customers. Ford learned and recovered from the fiasco, but the lesson was clearly not seen or learned by others. Regretfully, this is an ongoing area of potential conflict between engineers and corporations.

EXAMPLE 7.3

Boeing 737 Max 8

A current example remains a hotly debated topic, as airlines were forced to ground fleets of Boeing 737 Max 8 airliners after two crashes in 2018 that killed 346 people. Indonesia's aviation safety agency blamed a faulty sensor and problems with a Maneuvering Characteristics Augmentation System (MCAS) for the crash. Ethiopian officials have not published their final report, but an interim analysis also largely blamed the aircraft's design (Gates and Baker, 2019).

Once the National Transportation Safety Board started looking into the issue, problems with the sensor and the MCAS were suggested as a probable cause for the crashes. A series of internal discussions that reflected poorly on Boeing management were discovered. The sensor was initially designed to detect whether the nose is too high upon landing. If it is, the MCAS automatically pushes the nose down. The design of the system was to be dependent upon only one sensor, but it was found that if the sensor was faulty, the plane could not be controlled. An in-depth *Seattle Times* article notes that the "original version of MCAS . . . was activated only if two distinct sensors indicated such an extreme maneuver: a high angle of attack and a high G-force." (Gates and Baker, 2019).

continued

Most planes have two sensors. For the 737 Max, Boeing eliminated one of the sensors. The ultimate genesis for the problems can likely be traced to *costs versus engineering* issues since Boeing offered solutions to the one sensor issue for added costs that neither airline paid. Boeing revised the software, added a second sensor, and retested the flight simulators. After extensive testing (which uncovered even more issues), the 737 Max was cleared to fly in the United States in December 2020, nearly three years after the second crash.

7.4 LOBBYING/DISPARAGING COMMENTS

Every engineering firm must compete within the marketplace to procure work. In fact, every engineering firm, as in any business, must compete to gain clientele in order to survive. In this politically charged world, any effort can present challenges. One example is competing for public sector work and trying to lobby elected officials to involve their firm. The issues are many. The NSPE notes the following:

- **Canon 5a:** Engineers shall not falsify their qualifications or permit misrepresentation of their or their associates' qualifications
- **Professional Obligation 1a:** Engineers shall acknowledge their errors and shall not distort or alter the facts
- **Professional Obligation 6b:** Engineers shall not promote their own interest at the expense of the dignity and integrity of the profession

Work is tight. Money can be significant. Firms will attempt to overcome the perceived advantages of others. Certain states actually prohibit lobbying for work as well as selection based on costs (instead of based on qualifications), and some have local rules for codes of silence. Lobbying can easily get out of hand and lead to violations of all of these parameters. As the public becomes more aware of these activities, the common perception will be to focus on the business practices of developers and their ilk, as opposed to protecting the image of the profession. An example that is worth discussing follows.

EXAMPLE 7.4

Miami-Dade County Project Engineering

In 2014, the Miami-Dade Inspector General's Office issued a report regarding a solicitation for engineering services. The lobbying in response to the solicitation became fairly nasty and the Office of the Inspector General for Miami-Dade County investigated a series of accusations that were raised by the losing firm about the content of proposals and statements that had been made by another (winning) engineering firm as a part of the selection process. The losing team suggested that the winning team had disparaged them with 15 outright lies in order to secure the project. The project was worth $500 million, and a number of lawyers/lobbyists were involved.

Let us start with the legal issues. Florida Statute 287.055 states as its intent that "the agency shall negotiate a contract with the most qualified firm for professional services at compensation which the agency determines is fair, competitive, and reasonable." Note that the italicized wording from the statute below would suggest that the act of employing lobbyists and lawyers might violate:

Ch 287.055 (6) PROHIBITION AGAINST CONTINGENT FEES.—

(a) Each contract entered into by the agency for professional services must contain a prohibition against contingent fees as follows: "The architect (or registered surveyor and mapper or professional engineer, as applicable) warrants that he or she *has not employed or retained any company or person, other than a bona fide employee working solely for the architect (or registered surveyor and mapper, or professional engineer, as applicable) to solicit or secure this agreement* and that he or she has not paid or agreed to pay any person, company, corporation, individual, or firm, other than a bona fide employee working solely for the architect (or registered surveyor and mapper or professional engineer, as applicable) any fee, commission, percentage, gift, or other consideration contingent upon or resulting from the award or making of this agreement." For the breach or violation of this provision, the agency shall have the right to terminate the agreement without liability and, at its discretion, to deduct from the contract price, or otherwise

continued

recover, the full amount of such fee, commission, percentage, gift, or consideration. (*My emphasis added.*)

Looking at the ASCE code of ethics, the fundamental principles that may be in question within this concern are:

- Being honest and impartial and serving with fidelity the public, their employers, and clients
- Striving to increase the competence and prestige of the engineering profession's fundamental principles

Looking at the ASCE code of ethics, the fundamental canons are as follows:

1. Engineers shall hold paramount the safety, health, and welfare of the public and shall strive to comply with the principles of sustainable development in the performance of their professional duties.
2. Engineers shall perform services only in areas of their competence.
3. Engineers shall issue public statements only in an objective and truthful manner.
4. Engineers shall act in professional matters for each employer or client as faithful agents or trustees and shall avoid conflicts of interest.
5. Engineers shall build their professional reputation on the merit of their services and shall not compete unfairly with others.
6. Engineers shall act in such a manner as to uphold and enhance the honor, integrity, and dignity of the engineering profession and shall act with zero tolerance for bribery, fraud, and corruption.
7. Engineers shall continue their professional development throughout their careers and shall provide opportunities for the professional development of those engineers under their supervision.

The IG report suggested that the activities contrasted Canons 3, 5, and 6.

continued

Canon 3 states: "Engineers shall issue public statements only in an objective and truthful manner.

a. Engineers should endeavor to extend the public knowledge of engineering and sustainable development, and shall not participate in the dissemination of untrue, unfair, or exaggerated statements regarding engineering.
b. Engineers shall be objective and truthful in professional reports, statements, or testimony. They shall include all relevant and pertinent information in such reports, statements, or testimony.
c. Engineers, when serving as expert witnesses, shall express an engineering opinion only when it is founded upon adequate knowledge of the facts, upon a background of technical competence, and upon honest conviction.
d. Engineers shall issue no statements, criticisms, or arguments on engineering matters which are inspired or paid for by interested parties, unless they indicate on whose behalf the statements are made.
e. Engineers shall be dignified and modest in explaining their work and merit, and will avoid any act tending to promote their own interests at the expense of the integrity, honor, and dignity of the profession."

The Inspector General found that 3a, 3d, and 3e were specifically violated.

Canon 5 states: "Engineers shall build their professional reputation on the merit of their services and shall not compete unfairly with others.

a. Engineers shall not give, solicit, or receive either directly or indirectly, any political contribution, gratuity, or unlawful consideration in order to secure work, exclusive of securing salaried positions through employment agencies.
b. Engineers should negotiate contracts for professional services fairly and on the basis of demonstrated competence and qualifications for the type of professional service required.

continued

c. Engineers may request, propose, or accept professional commissions on a contingent basis only under circumstances in which their professional judgments would not be compromised.
d. Engineers shall not falsify or permit misrepresentation of their academic or professional qualifications or experience.
e. Engineers shall give proper credit for engineering work to those to whom credit is due and shall recognize the proprietary interests of others. Whenever possible, they shall name the person or persons who may be responsible for designs, inventions, writings, or other accomplishments.
f. Engineers may advertise professional services in a way that does not contain misleading language or is in any other manner derogatory to the dignity of the profession. Examples of permissible advertising are as follows:
    - Professional cards in recognized, dignified publications, and listings in rosters or directories published by responsible organizations, provided that the cards or listings are consistent in size and content and are in a section of the publication regularly devoted to such professional cards.
    - Brochures which factually describe experience, facilities, personnel and capacity to render service, providing they are not misleading with respect to the engineer's participation in projects described.
    - Display advertising in recognized dignified business and professional publications, providing it is factual and is not misleading with respect to the engineer's extent of participation in projects described.
    - A statement of the engineers' names or the name of the firm and statement of the type of service posted on projects for which they render services.
    - Preparation or authorization of descriptive articles for the lay or technical press, which are factual and dignified. Such articles shall not imply anything more than direct participation in the project described.

continued

- Permission by engineers for their names to be used in commercial advertisements, such as may be published by contractors, material suppliers, etc., only by means of a modest, dignified notation acknowledging the engineers' participation in the project described. Such permission shall not include public endorsement of proprietary products.

g. Engineers shall not maliciously or falsely, directly or indirectly, injure the professional reputation, prospects, practice, or employment of another engineer or indiscriminately criticize another's work.

h. Engineers shall not use equipment, supplies, laboratory, or office facilities of their employers to carry on outside private practice without the consent of their employers."

An argument could be made that many of these sections were violated by one or both parties. Setting aside 5a, of which there was much discussion, 5g may be the most egregious of the issues since this goes to the integrity of the profession.

Engineers have enough difficulty conveying what it is they do to the public (52% think they drive trains), creating confusion in the minds of many concerning our honesty and integrity. This would appear to be detrimental to the profession and of concern to the Board. The integrity of the profession is specifically what Canon 6 outlines:

Canon 6 states: "Engineers shall act in such a manner as to uphold and enhance the honor, integrity, and dignity of the engineering profession and shall act with zero tolerance for bribery, fraud, and corruption.

a. Engineers shall not knowingly engage in business or professional practices of a fraudulent, dishonest, or unethical nature.

b. Engineers shall be scrupulously honest in their control and spending of monies, and promote effective use of resources through open, honest, and impartial service with fidelity to the public, employers, associates, and clients.

c. Engineers shall act with zero tolerance for bribery, fraud, and corruption in all engineering or construction activities in which they are engaged.

continued

d. Engineers should be especially vigilant to maintain appropriate ethical behavior where payments of gratuities or bribes are institutionalized practices.
e. Engineers should strive for transparency in the procurement and execution of projects. Transparency includes disclosure of names, addresses, purposes, and fees or commissions paid for all agents facilitating projects.
f. Engineers should encourage the use of certifications specifying zero tolerance for bribery, fraud, and corruption in all contracts."

The Inspector General found that 6a, 6d, and 6e were not upheld in the process.

This was particularly upsetting, but as noted, it is not the first and is unlikely to be the last example of an attempt to subvert the solicitation processes. Regardless, this engagement was egregious enough that it attracted the attention of the State Inspector General.

The engineering community must seek guidance concerning any lobbying activities. The specifics of any situation or actions—whether intended or unintended—by others on behalf of engineers can be clearly in contrast to long-standing ethical principles that were established to protect the integrity of the profession.

7.5 THE GROUNDWATER CONUNDRUM

A United States Geological Survey (USGS) report (#1323 by Reilly et al., 2008) suggests that engineers have been over-drafting groundwater in many areas of the country. Figure 7.1 shows the difference between rainfall and evapotranspiration (ET). The light-colored areas indicate where the evapotranspiration rate is higher than the rainfall (meaning no net rainfall exists for crops and other purposes). It also indicates where groundwater is likely to be used, as confirmed by aerial views

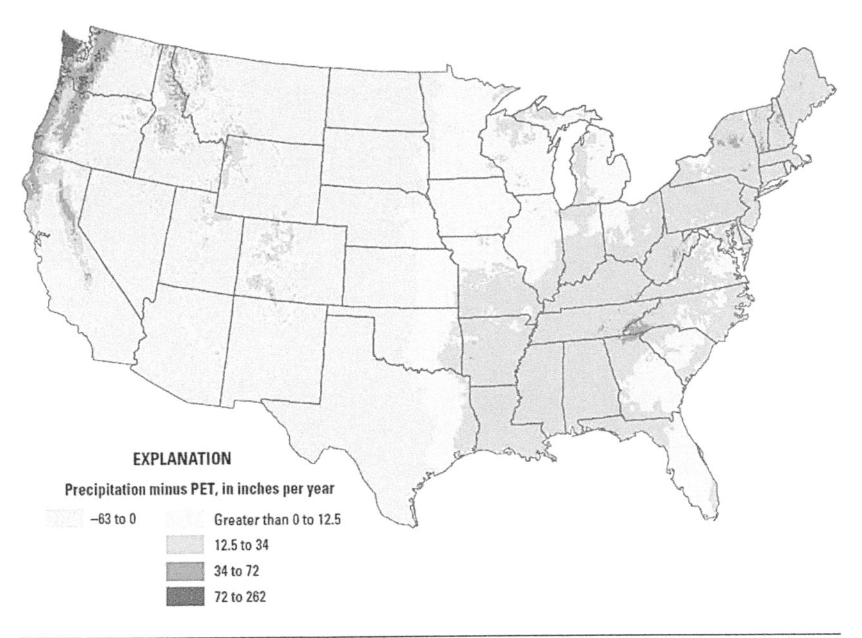

Figure 7.1 Difference between average annual precipitation and potential ET rates (*Source*: Reilly et al., 2008)

that show extensive irrigation or *crop circles* in many of these areas (see Figure 7.2).

Figure 7.3 shows that many of these areas never contain surplus amounts of water, thus high groundwater use is not a sustainable practice since recharge occurs at a very small rate. Reinforcing these conclusions is the data that are summarized in Figure 7.4, which shows the amount of water available for recharge throughout the United States.

Most areas have very little water available for recharge. Figure 7.5 combines regional water-level declines and local water-level declines for changes over the last 40 years throughout the United States. The light-colored regions indicate areas in excess of 500 square miles that have water-level decline in excess of 40 feet in at least one confined aquifer since predevelopment or in excess of 25 feet of decline in unconfined aquifers since predevelopment.

Figure 7.2 Aerial view of *crop circles*; irrigated areas from groundwater pumpage in the arid West

The dark dots are wells in the USGS National Water Information System database where the measured water-level difference over time is equal to or greater than 40 feet. These reflect areas noted in previous maps; areas where there are already indications that water supplies are insufficient to provide the full needs of the community (Reilly et al., 2008; Bloetscher et al., 2009; Bloetscher and Muniz, 2012). The lowering water levels are viewed by state agencies as an indication that recharge is generally overestimated, giving a false picture of water availability. If aquifer use declines year after year, it is not necessarily caused by drought—it may be being mined. The continued withdrawal of water from that source results in a permanent loss of the resource in the long term.

Reilly et al. (2008) found that the loss of groundwater supplies in many areas could be potentially catastrophic, affecting the economic viability of communities and disrupting lives and ecological viability.

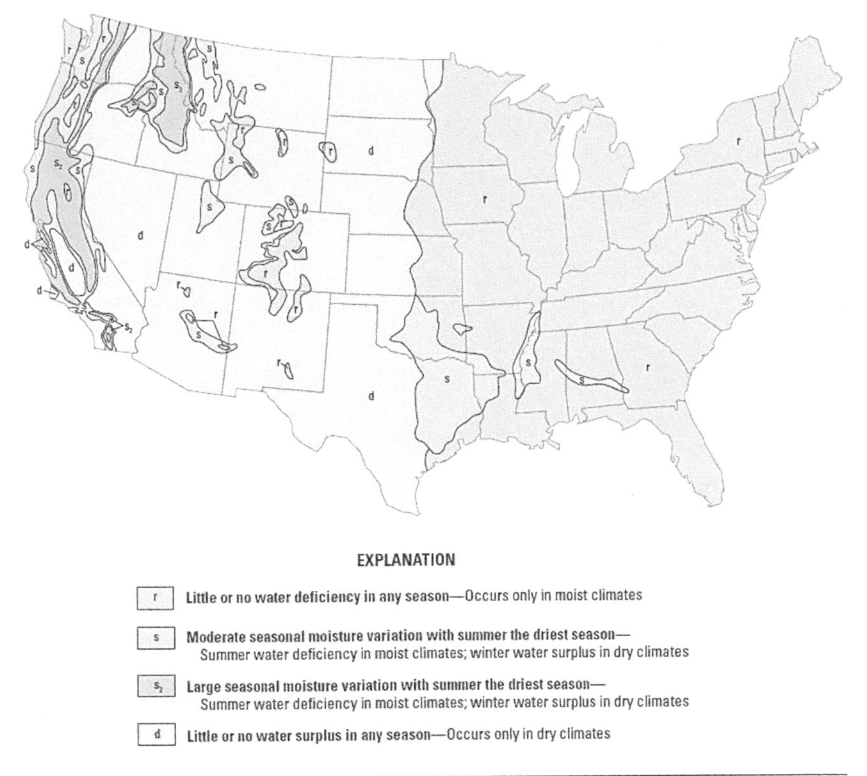

Figure 7.3 Water deficit areas (*Source*: Reilly et al., 2008)

DOE/NETL (2009) suggested that by 2025, the most vulnerable areas for water shortages will be among these fast-growing areas: Charlotte, NC; Chicago, IL; Queens, NY; Atlanta, GA; Dallas, TX; Houston, TX; San Antonio, TX; and San Francisco, CA. Immediately behind these areas are Denver, CO; Las Vegas, NV; St. Paul, MN; and Portland, OR. Some areas have more options than others but drilling deeper is not a long-term solution.

Drawing groundwater at greater depths typically provides poorer water quality as a result of waters having been in contact with the rock formation for longer periods. Minerals within the rocks will dissolve into the water. Therefore, if deeper, but harder (and saltier) water is withdrawn, additional power will be required to further treat the

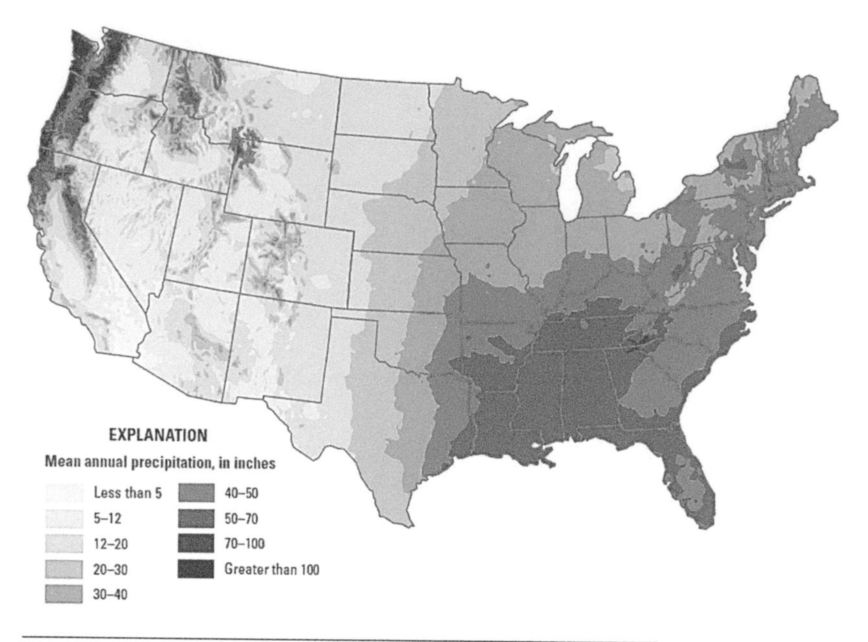

Figure 7.4 Water available for recharge throughout the United States; note most areas are very low (*Source*: Reilly et al., 2008)

limited, lower quality supplies. Therefore, while some deep aquifers may be prolific, the quality of water obtained from a well may not be desirable or even potable without substantial amounts of treatment.

Deeper aquifers are generally confined, and therefore do not significantly recharge on the local level. In many cases, going deeper does not provide a solution, so lowering groundwater levels is a not-too-distant future problem, especially for small communities. Groundwater has served as a viable solution within small utility and agricultural applications. Lowering levels of groundwater puts many of the over 40,000 groundwater systems (which serve nearly 100 million Americans and millions of acres of agriculture) at risk. This mirrors conditions within communities in similar positions, as seen in the eastern Carolinas during the mid-1990s.

For instance, let us say a client hires an engineer to find a water supply for his activity—a typical request. The location of the site is within

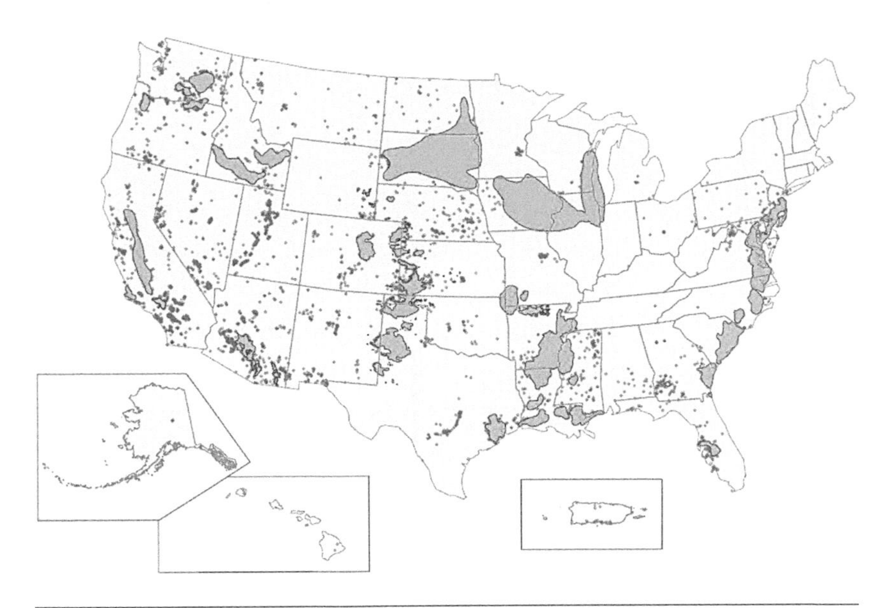

Figure 7.5 Water-level declines. Red regions indicate areas in excess of 500 square miles that have water-level decline in excess of 40 feet in at least one confined aquifer since predevelopment or in excess of 25 feet of decline in unconfined aquifers since predevelopment. Blue dots are wells in the USGS National Water Information System database where the measured water-level difference over time is equal to or greater than 40 feet. (*Source*: Reilly et al., 2008)

an area where groundwater is decreasing at a rapid rate. What, realistically, are your options? NSPE Canon 1 states "Hold paramount the safety, health, and welfare of the public." If the groundwater supplies are decreasing, this canon cannot be upheld by increasing withdrawals from the water supply. But that is not what the client wants. The engineer cannot *create* a model that shows groundwater is available when it clearly is not—at least not without violating Canon 5: "avoid deceptive acts." Technical opinions must "be founded upon knowledge of the facts and competence in the subject matter."

There are times when engineers must acknowledge that some projects will not be successful. If water is overcommitted in a watershed, then drilling more or deeper wells is not in the public interest. The concern for public welfare is not met. The engineer must advise the client of this fact. That may not prolong the engineer's involvement on-site, but it protects the public health, safety, and welfare while upholding the dignity of the profession.

Recent revisions to codes of ethics for engineers have included sustainability. Sustainable groundwater will be a twenty-first century issue for the public, engineers, industry, water utilities, and agriculture. Competition will complicate the discovery of solutions. Given that many watersheds are overcommitted, prioritizing water use under such scenarios must include concerns for water quality needs, so that poor quality waters can be prioritized for lower priority usage such as agriculture. However, the water quality needs of ecosystems have been historically discounted, as have intrinsic economic values. As a result, practices such as large-scale clearing and filling of coastal wetlands, bayous, and mangroves have subjected coastal areas to damage from storm events and sea-level rise.

Ultimately, securing reliable water supplies for future generations is critical in the face of changes in climatic patterns. Water supplies can become more reliable and sustainable through a comprehensive approach to water planning, which might include alternative water sources and future infrastructure over long-term trends. Systems need to adapt to changing conditions, as past long-term prognoses are expected to be altered because of climate impacts.

Implicit in the evaluation of these sustainability concepts is the uncertainty and the reversibility of shortages in natural resources. Sustainability should be defined as being continuous and whether that uninterrupted supply is inclusive. Inclusivity should account for multiple objectives, lest money or resources be wasted in competition between these factors (Popp et al., 2001).

The NSPE code of ethics states that engineers must "hold paramount the safety, health, and welfare of the public." That is not very helpful without realizing that the long-term sustainability of society is part of this canon. As with groundwater withdrawals, telling the truth and issuing "public statements only in an objective and truthful manner" will be expected in order to comply with Canon 4.

Ultimately, this conundrum suggests that we need to maintain a dialogue to better manage groundwater resources domestically. But can we achieve this without all of the legal and political constraints that currently work against protecting our nation's groundwater supplies? While clients do not hire engineers *to not find water*, the public welfare suggests that engineers—as members of a profession—are obligated to protect the citizenry. This means protecting water supplies for all, not just clients.

Clearly that will not make everyone happy, and may make a lot of people very unhappy, but better to make those decisions now, than make them in 20 or 30 years when the groundwater runs out. The goal of elected officials and developers is to continue to build to attract more people and business. Engineers have a responsibility to protect the public's interest. As a result, the growth objectives of many elected officials are expected to be challenged in many environments where the water supplies are relatively finite or fixed. Inevitably, the supply will be exceeded by local demands (the opposite of *sustainability* from a water resource perspective).

For many places, groundwater should probably not be the primary source, and should be a backup plan only. This is a difficult analysis for many to understand. So, let us look at an example in order to understand how one might address making recommendations.

7.5.1 Klamath River Basin

A Government Accountability Office (GAO) report roundly criticized the local administration and Washington bureaucrats for providing irrigation diversions during a drought that wiped out salmon runs in the Klamath River Basin during 2000 and 2001 (GAO 2005). In the report it was seen that the economic impact associated with the Klamath River Basin included a $4.5 billion per year salmon fishery industry that had been reduced to $50 million per year as a result of agricultural diversions (that were said to produce an economic benefit of only $200 million per year). A diversion in any given year would potentially impact the supply of salmon for years since salmon return at specific intervals (4–6 years from birth).

Also noted was that recreational opportunities (mostly concerned with fishing and bird watching in large wildlife sanctuaries) was the fastest growing segment of the economy in the area—$800 million per year and expected to grow at a rate of $50 million per year. If salmon fishing was to provide reliable economic opportunities, the potential within this sector would dwarf all others (GAO, 2005). But no evaluation of economic impacts was made.

The water supplies are overcommitted. Some, but not all, of the demands can be met. Which are prioritized? *The dilemma could be simply stated as agriculture versus ecosystem versus economics versus politics.* Prioritizing water use under such scenarios should include water quality considerations, so that lower quality waters can be prioritized for usage needs such as agriculture. In the Klamath case, an economic argument could easily be made that fish and tourism were far higher priorities than the agricultural interests that received the water. However, that is not where the votes were.

Any study of sustainability should consider whether sustainability is so inclusive as to meet multiple objectives, lest money or resources be wasted in competition between goals (Popp, et al., 2001). Depleted ecological stocks cannot be sustained and conserving natural capital does not always imply absolute protection against depletion.

Compensatory mitigation is one replacement strategy that has been tried and pursued in an attempt to mitigate impacts to ecological systems (National Research Council, 1992). Therefore, questions remain whether resources should be sustained as they currently exist and how to balance the services provided.

In such cases, the value of the best alternative is not always provided. For example, if a wilderness area is timbered, would it cease to be a productive fishery? Is this a net positive or net negative to the sustainability of the basin? Would the opportunity cost of the timber alternative equal the value of the fishery lost? Or, if you choose to build a hydropower plant, as opposed to a coal-fired plant, the optimal design choice might be to avoid the cost to provide the mandated pollution equipment needed to clean the air. There is a net positive to the project.

7.6 RISK ASSESSMENT: IS FRACKING DESIRABLE?

How do engineers evaluate something seen as vitally needed but controversial? An example would be fracking and the relative risk of pollution of drinking water supplies from these types of operations. Risk assessments are one answer, but these are probabilistic exercises with definite amounts of uncertainty, something engineers prefer not to deal with.

The word *risk* raises concern within the public because it is often related to a public health issue. However, the definition of the term and its connotations are unclear to most people. *Risk* is defined as the *probability of occurrence of an event*—in this case, the withdrawal of a contaminant in drinking water introduced into the aquifer by a fracking project. However, attempting to quantify and measure such risks leads to more questions by the public as to what type of risk is acceptable and what is not. The U.S. Environmental Protection Agency has defined acceptable risk for drinking water purposes as a one in a million chance of illness per year and a lifetime risk of 1:10,000. At what point is the risk greatest? Should P.E.s be involved?

In the case of fracking, there are three phases of development: investigation, well drilling, and operation, which opens up four potential risk exposure pathways. *Well drilling* presents the risk potential for leakage at the surface from operations or into the ground because of the fracking protocol. Most fracking sites maintain constant operation once placed into service, although periodic maintenance or refracking may be required to ensure that yield values remain high enough to provide a profit from the operation of the well. Proper disposal of wastewater is also necessary.

Continuous operations can provide an exaggerated perception of the operation. These wells will not physically be able to operate at all times due to operational and mechanical maintenance and disruptions, and the volume withdrawn can be expected to vary with time. The quantity of water demanded for the various phases may be inconsistent as well. The start-up process will require the most water, which must be drawn from a reliable source, and disposed of or recycled after its use. A lesser amount of water is used during operations, but long-term migration can occur.

So, what if your client wants to design fracking wells? What if the job site is located in Oklahoma (or Ohio) where the number of earthquakes has increased dramatically as an apparent result of injecting the wastewater into the subsurface? What are the options? NSPE Canon 1 dictates that an engineer must hold public health, safety, and welfare paramount. An engineer might conduct a risk assessment of the hazards posed by these chemicals to provide insight into the potential public impact from waters that might contain such chemicals.

This risk assessment should involve professionals and scientists who can develop standards of practice and protocols for fracking well usage. A desired outcome would specifically define the risk of using potentially contaminated waters for public purposes.

Unfortunately, risk assessment has only been implemented to a limited extent when judging the risks of chemical contaminants. An engineer would need to judicially use public statements only as a way to provide objective and truthful communication to convey this information. This requirement behooves engineers to conduct themselves honorably, responsibly, and ethically.

Therefore, the question is—when is safety the paramount consideration? Is this project safe? Can we make it safe? Noting that gas is necessary for society to be efficient and operational, where does the public welfare lie? This answer should incorporate concern for the public welfare everywhere. This is why consideration of potential impacts is important when assessing risk. Decisions must be based on whether the public health, safety, and welfare are protected.

7.7 ARTIFICIAL INTELLIGENCE (AI)

Many people would like to believe that AI will lead to cool technology such as automated robots and the like, as seen in Hollywood movies. That is a misrepresentation of what AI is and what its potential may contain. AI is simply the use of computer software to predict the most likely event based on inputting many prior events, some created by the software itself. Terms like neural networks, predictive Bayesian statistical methods, bootstrapping methods, etc., are used to describe how algorithms are created to develop AI programs. For example, AI can be trained to play chess and figure out the most likely set of moves an opponent will make, while designing effective countermoves.

What AI does not do is predict a *black swan*. A black swan is a phenomenon described by Nassim Nicholas Taleb in his book with the same name (Taleb, 2008). These events represent extreme outliers that are unlikely to be considered within the realm of probabilities. They can also be seen to have disruptive effects that may influence behavior going forward. COVID-19 might be such an event, as might be the Great Recession—or AI.

Applications for AI technology are many. The business field has extensively relied on AI within the world of finance to improve efficiency and limit risk. Medical personnel use AI programs for diagnosis and to find treatment solutions. The criminal justice system uses AI for predicting crime patterns, recidivism, and to determine sentencing guidelines. Hydrologic engineers use AI algorithms to forecast water levels, flood patterns, and rainfall.

In all cases the AI computer program will use prior information or created prior information to predict most likely probabilities; but AI will not predict the actual result. AI will create a series of results with probabilities for each. That is one underlying reason why criminal justice programs might appear to be biased. It is also one underlying reason why lending programs appear to be unfair.

Jamie Dimon, CEO of JPMorgan Chase Bank (who has made extensive use of AI for decision making and productivity improvements), recognized this bias during a *60 Minutes* program recently, stating that input that attempts to replicate prior information can produce results that can be perceived as unfair. For example, all of his AI algorithms create results that are less likely to endorse making loans to women and minorities. He notes that the software typically inherits the biases of the programmer and this type of data is often used to *train* the AI algorithm. As a result, these programming procedures remove a humane aspect of decision making, resulting in ongoing bias. As a result, minorities, women, and young people are less likely to get loans than white males that are middle aged and older. JPMorgan Chase is investing heavily in Detroit to address bias that the firm believes has occurred over the years.

Is AI a solution for vehicle accidents? Self-driving trucks are already on the road. There are 1.3 million long-haul truckers at this time. Over-the-road truck driving is considered among the highest wage jobs for those without a college degree. It also accounts for 10 percent of all economic activity. The argument for autonomous vehicles is that there are no human drivers who might suffer fatigue so the incidence of accidents would be lessened. They are also cheaper to operate (no wages and benefits to be paid to a human driver). So, what happens to the truckers? Are they the next iteration of laid off workers like the Concord Coach makers or sanitation shovelers? There is a downside to innovation, and it is obvious AI invariably has a human impact.

A major push exists within Silicon Valley tech firms and automakers to develop self-driving cars, but self-driving cars cannot be expected to operate to perfection because the AI program cannot know all of the solutions. Self-driving cars will still be involved in accidents because no computer program can predict highly unlikely events accurately. The question will be who is responsible for such accidents. How does the car decide between potentially injuring five kids who are crossing the roadway or the vehicle's passenger if only one of those two options is possible? (This is a classic philosophy discussion since it involves an example of street cars from another era.)

AI can be counterproductive. Meredith Broussard (2018) notes that algorithms used within criminal sentencing have a major bias against young male minorities. While it is illegal to specifically consider race, zip codes can be utilized to make projections concerning ethnicity, income, and other factors that would indicate lower income and minority populations. The bias in these programs suggests people from certain zip codes are likely to commit future crimes. So, the software increases penalties against these residents, even for first-time offenders. In contrast, those offenders who live in areas that are economically defined as middle and upper class yet commit the same types of crimes typically receive lesser sentences than those of the underclass. Can the use of AI improve social equity? The initial answer, at this time, is *no*.

In *Weapons of Math Destruction*, Cathy O'Neil (2018) notes that there are few regulations or constraints on the application of AI techniques. As a result, aside from banks and governments, predatory AI programs subject many lower income people to disparate treatment based on algorithms about who they are and where they live. Loans, insurance, and health options are among the AI treatments that are subject to this type of activity. AI uses your unused cellphone to robodial others. Have you noticed that?

The author, Chris Wylie, in his book *Mindf*ck*, concerning the Cambridge Analytica (CA) scandal that he created, demonstrated how AI algorithms can be used to adjust political views, connect like-minded persons, and additionally permit foreign entities to send messages that support their favored politicians (or against those that are not) in order to potentially destabilize governments.

Wylie participated in such efforts in Canada, Great Britain, and the United States before 2017. He acknowledged Russian efforts to impact the election results in the United States. While CA no longer exists, these AI programs are still at work and many of his compatriots at CA are working for political parties, the military, and commercial venues. What they do works for all categories—they harvest data from online platforms and tailor the online experience of individuals to maximize the number of clicks a person might make. This is why two neighbors get ads for paper towels, toilet paper, and napkins, but one gets ads for baby products, and the other receives promotions for sports drinks.

The AI program knows what people buy and sends ads for those products. How? AI algorithms gather data on the web and send personalized information that is likely to be opened and read. It will not send information that users would be unlikely to open. Obviously, AI can provide great benefits, but nefarious aspects exist. Wylie notes that AI algorithms were used in efforts to destabilize society during the 2016 U.S. election, as well as in Great Britain during Brexit and in Nigeria, Trinidad, and other places (and apparently continue to do so).

Data mining simply invokes AI algorithms that seek patterns to focus marketing efforts upon targeted entities. The practice is typically deemed to be effective because the web is cheaper to utilize than the mail. Hence, no two people are likely to consider the same things, and they never see the same information considered by others. This opens the argument of *alternative facts* (which, by definition, are not possible).

The reliability, factualness, and intent of the content is not considered, it is only traffic to be measured to the site. That portends poorly for a person to be able to facilitate constructive conversation and find common ground, largely in part since the information is often perceived as disconnected.

Wylie (2019) notes that computer programmers do not generally have a code of ethics, while engineers do. This is a murky area. It may be one reason why governments might attempt to limit predatory programs at some point. This may lead Facebook and Twitter to shut down an individual's account. Selling products is one thing, selling fake facts is different. Journalism is indisputably the lowest paying profession that requires a college-level education, but it purports to have a code of ethics that goes back hundreds of years, while access to and use of the internet does not. It could be expected that governments could extend that type of requirement to the internet.

The code of ethics for NSPE (2020) states that engineers must "Hold paramount the safety, health, and welfare of the public" and to act honorably. Professional Obligation 8 notes that "Engineers shall accept personal responsibility for their professional activities, ..." So, the issues with accidents from tools and equipment are a challenge (i.e., Boeing 737 Max 8). But the canons and obligations do not address the issue of fairness, except that it might be attributed to general welfare. ASCE Canon 8, however, does state that "in all matters related to their profession, treat all persons fairly and encourage equitable participation without regard to gender or gender identity, race, national origin, ethnicity, religion, age, sexual orientation, disability, political affiliation, or family, marital, or economic status." The application of AI could be difficult to quantify in these humanistic areas.

The application of AI is a complex consideration, and it can be expected that engineers will develop many tools, projects, and initiatives that will incorporate the use of AI in the future. One must understand the potential downside to any AI initiative. An obvious realization might be that computers do not consider the human element, but only the results programmed as outcomes. What are those outcomes? We should know because we will have to live with these outcomes.

7.8 CONTRIBUTING TO POLITICIANS

ASCE strongly supports the involvement of civil engineers in the legislative and regulatory decision-making processes. However, Hoke (2016) notes that in one 1971 case, ASCE members made campaign contributions to public officials while benefitting from public contracts. The donors maintained that their donations reflected their *civic duty* to play a role in the election process. ASCE questioned the public good associated with such contributions, creating a five-point guideline submitted to ASCE's Board of Direction which was approved in April 1971—it reads as follows (ASCE, 2012):

1. "Every engineer has the same rights as other citizens to participate in the political process, to contribute his [or her] time and money to political campaigns, [and] to attempt to influence legislation, executive decisions, and appointments. However, offering to pay, paying, or accepting either directly or indirectly, any gift, bribe, or other consideration to influence the award of professional work, shall be considered unethical conduct.
2. Every engineer making a political contribution should do so publicly in his own name, and as an individual citizen.
3. Every engineer has the right to refuse to contribute to any organization, political campaign, or candidate for public office.
4. Every engineer has the duty to hold public officials accountable for their actions, particularly on those matters in which the engineer has specialized professional competence.
5. An engineer who makes a direct or indirect political contribution in any form under circumstances that are related to his selection for professional work shall be (a) subject to disciplinary action by the Society, and if appropriate, (b) reported to the public authorities."

7.9 SUMMARY

As can be seen from the more complicated examples in this chapter, not everything is easy to understand. It is why we tried to clear the fog in Chapter 2. But there are many potential pitfalls along the way. Going all the way back to Mantell (1964), from an ethical guidance perspective, engineers must consider the moral imperative: "If everybody does X, does society function?"

Identify whether the engineer's actions are ethical and define the duties that the engineer must comply with. Once that threshold of criteria is recognized, Johanssen (2009) suggests the following problem-solving steps:

- "State the problem: Clearly define what the ethical engineering problem is.
- Get the facts: Obtain all relevant facts to the matter (i.e., the different moral viewpoints) and then analyze them all.
- Identify and defend competing moral viewpoints: Analyze the pros and cons of different moral viewpoints and pick the best course of action.
- Come up with a course of action: Pick the best course of action and answer all unanswered questions.
- Qualify the course of action: Back up the course of action with facts or statistics."

Ethics cases rarely have easy answers, but the NSPE's Board of Ethical Reviews examines nearly 500 advisory opinions.

Engineers should, as noted at the start of the chapter, be cognizant of their role as agents of social change and ensure that their efforts improve the general health, safety, and welfare of society. That will involve leadership and tough decision making.

REFERENCES

Abkowitz, M.K. (2008). *Operational Risk Management.* Wiley. Hoboken, NJ.

Baker, B.W. (2008). Engineering Ethics: Applications and Responsibilities. In *Engineering Ethics: Concepts, Viewpoints, Cases and Codes.* Lubbock, TX. National Institute for Engineering Ethics. pp. 49–65.

Bloetscher, Frederick. (1999). Looking for Quality on an Engineering Consultant. *American City and County*. December 1999, p. 36.

———. (1999a). What You Should Expect from Your Consulting Professionals (and How to Evaluate Them to Get It). *Water Engineering and Management*. October 1999. pp. 24–27.

———. (2009). *Water Basics for Decision Makers: What Local Officials Need to Know about Water and Wastewater Systems*. America Water Works Association. Denver, CO.

———. (2011). *Utility Management for Water and Wastewater Operators*. AWWA. Denver, CO.

Bloetscher, F. and D.E. Meeroff. (2015). *Practical Concepts for Capstone Design Engineering*. J. Ross Publishing. Plantation, FL.

Bloetscher, F., D.E. Meeroff, and B.N. Heimlich. (2009). *Improving the Resilience of a Municipal Water Utility against the Likely Impacts of Climate Change, A Case Study: City of Pompano Beach Water Utility*. Florida Atlantic University. November 2009. http://www.ces.fau.edu/files/projects/climate_change/PompanoBeachWater_Case Study.pdf.

Bloetscher, F. and A. Muniz. (2012). Where Is the Power to Treat all the Water? Potential Utility Driven Solutions to the Coming Power-Water Conflict. *Florida Water Resource Journal*. Vol. 64:3, pp. 32–46.

Bomey, Nathan. (2014). *GM recalls 1.3 million vehicles for power steering defect*. https://www.staugustine.com/article/20140401/NEWS/304019960#:~:text=General%20Motors%20Co.,cost%20it%20about%20%24750%20million.&text=GM%20said%20it%20would%20make%20free%20repairs%20on%20the%20vehicles. Accessed 10/14/2020.

Broussard, M. (2018). *Artificial Unintelligence*. MIT Press. Cambridge, MA.

Brown, S. (2020). *A new study measures the actual impact of robots on jobs. It's significant*. https://mitsloan.mit.edu/ideas-made-to-matter/a-new-study-measures-actual-impact-robots-jobs-its-significant.

DOE/NETL. (2009). Use of Non-traditional Water for Power Plant Applications; An Overview of DOE/NETL R&D Efforts. DOE/NETL-311/040609. Department of Energy. National Energy Technology Laboratory. Pittsburgh, PA.

Florman, Samuel. (1983). Commentary. In *The DC-10.* John H. Fielder and Douglas Birsch (eds.). State University of New York Press. Albany, NY.

GAO. (2005). *Klamath River Basin: Reclamation Met Its Water Bank Obligations but Information Provided to Water Bank Stakeholders Could Be Improved.* GAO-05-283. GAO. Washington, DC.

Gates, D. and M. Baker. (2019). Engineers Say Boeing Pushed to Limit Safety Testing in Race to Certify Planes, Including 737 MAX. *The Seattle Times.*

Gioia, Dennis A. (May 1992). Pinto Fires and Personal Ethics: a Script Analysis of Missed Opportunities. *Journal of Business Ethics, 11*(5–6): 379–389. doi:10.1007/BF00870550. S2CID 189918233.

Grush, E.S. and C.S. Saunby. (1973). *Fatalities Associated with Crash Induced Fuel Leakage and Fires.* Internal Ford memo. https://www.autosafety.org/wp-content/uploads/import/phpq3mJ7F_FordMemo.pdf. Accessed 10/14/2020.

Herkert, Joseph R. (2001). Future Directions in Engineering Ethics Research: Microethics, Macroethics and the Role of Professional Societies. *Science and Engineering Ethics.* VII (3): 403–414.

Hoke, T. (2016). *5 Ethical Gray Areas in Civil Engineering.* https://source.asce.org/five-ethical-gray-areas-in-civil-engineering/.

Hoyle, Andrew/CNET. (2019). *What Caused the Crashes*? https://www.cnet.com/news/boeing-737-max-8-all-about-the-aircraft-flight-ban-and-investigations/. Accessed 10/14/2020.

Jonassen, D.H., D. Shen, R.M. Marra, Young-Hoan Cho, J.L. Lo, V.K. Lohani. (2009). Engaging and Supporting Problem Solving in Engineering Ethics. *Journal of Engineering Education, 98*(3): 235–254. doi: 10.1002/j.2168-9830.2009.tb01022.x.

Jonsen, Albert R. and Stephen E. Toulmin. (1988). *The Abuse of Casuistry.* University of California Press. Berkeley, CA.

Kant, Immanuel. (1785). First Section: Transition from the Common Rational Knowledge of Morals to the Philosophical. *Groundwork of the Metaphysic of Morals.*

Levine Law. (2019). The GM Recall: Did the Auto Industry Forget the Lessons of the Pinto? https://www.mydenveraccidentlawfirm.com/news-resources/the-gm-recall-did-the-auto-industry-forget-the-lessons-of-the-pinto/. Accessed 10/14/2020.

Lienert, P. and M. Thompson. (2014). GM Avoided Defective Switch Redesign in 2005 to Save a Dollar Each. https://www.reuters.com/article/us-gm-recall-delphi/gm-avoided-defective-switch-redesign-in-2005-to-save-a-dollar-each-idUSBREA3105R20140402. Accessed 10/14/2020.

Mantell, M.L. (1964). *Ethics and Professionalism in Engineering.* Collier-MacMillan Ltd. London, UK.

Marcy, W. and J.B. Rathburn. (n.d.). *Engineering Ethics and Its Impact on Society.* https://ethicalengineer.ttu.edu/articles/engineering-ethics-and-its-impact-on-society. Accessed 2/14/2021.

Marcy, W. and R. Burgess. (2015). *Social Impact Analysis, Lecture ENGR 2392 Engineering Ethics and Its Impact on Society.* Texas Tech University. Lubbock, TX.

National Research Council. (1992). *Restoration of Aquatic Ecosystems: Science, Technology, and Public Policy.* National Academy Press, Washington, DC.

National Research Council. (2004). *Assigning Economic Value to Natural Resources.* National Academy Press. Washington, DC.

NCEES. (2020). Squared, National Council of Examiners for Engineering and Surveying. https://ncees.org/wp-content/uploads/Squared-2019.pdf.

Nyugen, D. (2013). Engineering Ethics. https://sites.tufts.edu/eeseniordesignhandbook/2013/engineering-ethics-2/. Accessed 10/14/2020.

O'Neil, C. (2016). *Weapons of Math Destruction.* Crown, Random House, LLC. New York, NY.

Popp, J., D. Hoag, and D.E. Hyatt. (2001). Sustainability Indices with Multiple Objectives. *Ecological Indicators.* Vol. 1. pp. 37–47.

Reilly, Thomas E., Kevin F. Dennehy, William M. Alley, and William L. Cunningham. (2008). Ground-Water Availability in the United States. *USGS Circular 1323.* USGS. Reston, VA.

Schrader-Frechette, K. (1985). *Science Policy, Ethics, and Economic Methodology.* Kluwer. Boston, MA.

Schrader-Frechette, K. (1991). *Risk and Rationality: Philosophical Foundations for Populist Reforms.* University of California Press. Berkeley, CA.

Taleb, N. (2008). *The Black Swan: The Impact of the Highly Improbable.* Random House, LLC. New York, NY.

Tattershall, S. (n.d.). https://www.quora.com/What-are-some-examples-of-engineering-ethics.

Wylie, C. (2019). *Mindf*ck.* Random House, LLC. New York, NY.

This book has free material available for download from the Web Added Value™ resource center at *www.jrosspub.com*

CHAPTER 8

ECONOMICS AND ENGINEERING—THE NEED FOR LEADERSHIP

Someone once asked Tina Turner to sing a song that was nice and easy. Tina replied, "Nothing that I have known in my life was *nice and easy*." They made a movie about all of that. Economics is like a Tina Turner song—truthful and real. Nothing is nice and easy.

If people are on the wrong side of economics, the effort to rise is a nearly insurmountable challenge. While engineers are not economists, engineers need to have enough knowledge of economics to understand the impact that their decisions or recommendations may have on others. This is part of the public health, safety, and welfare obligation. And to that end, there are a number of instances where economic hardships among other parties may be impactful.

Examples include the inevitable gentrification of neighborhoods, relocation versus rebuilding factories, artificial intelligence, stadium design, and social issues which are often transformed through computerization. Engineering professionals will be asked to be involved in *solutions* to these problems. Realizing the limitations of technology and its capabilities to impact economic and societal conditions is a critical responsibility that engineers must meet in order to protect the public health, safety, and welfare.

Overcoming the challenges posed in the prior chapters requires leadership and the willingness and capability to make tough decisions. Hence, a discussion concerning leadership is appropriate. The quality of leadership can only be defined in hindsight, i.e., our actions can only be judged after the effects are understood, so it is often difficult to gauge a leader's quality during the time of the actual occurrence of an event.

Further, questions remain on how to measure the value added of leadership, which could include improvements in the organization's financial state, system resiliency, overall reliability, or service levels. Among the desired outcomes is the capability to identify the qualities that make effective industry leaders recognizable, as well as to gain a clear perception of the barriers they might face. While envisioning the potential and probable leaders of tomorrow, it remains difficult to define exactly what to emulate.

LEARNING OBJECTIVES

- Defining leadership
- Understanding perceptions in the profession
- Setting a vision for tomorrow

8.1 DEFINING LEADERSHIP

Leadership is both a short-term and a long-term concern. For the long term, the focus must be placed on how the decisions being made today will impact the course of the organization and society in the future. One thing many people may not understand is that while we all live in the moment, it is how we will perceive the actions of ourselves and others in retrospect afterward that defines leadership.

As mentioned at a public official's workshop at the Annual Conference & Exposition, ACE18, we tend to remember those who implemented a vision, navigated through a crisis, or left behind a legacy that demonstrated that the system was better after the fact. This is a legacy leadership issue that should be considered by elected and appointed officials alike, as well as employees of any organization. A more difficult challenge is to determine the value added of any decision, since monetary value is not the only means to quantify this measure.

Leaders are generally more visible during times of change and crises. They are singularly evaluated upon a vision toward implementing changes in direction, as opposed to managers and technical experts who deal more with day-to-day business concerns. That is a challenge for engineers who are often caught up with day-to-day issues and might fail to focus on leadership for the future.

The challenge before the engineering industry currently is to seek out tomorrow's leaders now and then put them in positions to succeed. Unfortunately, the engineering profession must compete with other fields that may already have recognizable leaders who can attract attention and funding for their organizations. In addition to helping the industry evolve, leaders in the engineering field need an appreciation of public health protection and community sustainability, along with an understanding of infrastructure renewal and reliability.

When one questions leadership, there are many answers, as indicated by the more than 10,000 publications on leadership, each indicating that it has a unique and singular perspective on answers, solutions, tools, or skills. It is hard to define leadership because it comes in many forms and is often specific to the approach to a situation. A quarterback who is a great leader on the field might not be the best choice to lead the reorganization of a major corporation. Both positions require leadership, but the skill sets required for each position is situational.

Because one cannot define what skill set is needed for every situation, the common tendency is to look at examples of people who are leaders or who have exhibited leadership in the past and try to draw from their experience; thereby replicating what made them a leader. It was not clear that Abraham Lincoln would be remembered as one of the greatest leaders in American history at the time that he was president. Anyone who reads accounts of his presidency would realize that the early years were marked with indecision and backtracking. Most of that is forgotten in lieu of the ultimate results. Results matter and conveying the proper message is required.

Leadership and ethics are interrelated; both are recognized more when they are absent—less so in the moment. An individual cannot be a leader if he or she is perceived to be unethical. If someone offends the sensibilities of any individual, they usually will not follow that person (although other morally lacking people might—e.g., Hitler).

Another conundrum concerning effective leadership is that people tend not to understand what it is. For example, leadership and management are not the same thing. Managers work with resources to maintain and sustain a process. Leaders generally set and often alter the goals. Leaders are generally associated more with change and crises, with vision, and with changes in direction, than managers who deal more with day-to-day business focus.

There are many good managers in the engineering profession, but managers, even great ones, are rarely acknowledged. Leaders are different—they can be inspirational, charismatic, visionary, innovative, and challenging—and they may be successful or unsuccessful. They put their name on what they do. People recognize them.

Leadership and individual leaders can be evaluated by those who follow them. An effective leader will also seek to bring in people to fill any gaps. That means leaders will hire the best people they can, without worrying about whether they are vying for some future position. In some cases, the strategy must be adjusted based on the skill sets of people on hand.

Leaders must clearly communicate their vision, or risk losing their position of authority. Next, they will need to involve those who buy into that vision, or they will not effectively lead. Leaders must have confidence in their abilities and strive to make everyone in the organization better by not putting them in a position to fail.

When presented with a challenge, how the leader attacks it and how they marshal the resources to succeed is what is observable. As a result, leaders may exist at every level of the organization. The challenge is to identify them and ensure that they are supported and placed in a position to succeed. When the leader finds the right mix, success follows.

Lincoln found this during the Civil War when he spent time with the troops. He communicated his vision to the soldiers, and he expressed his appreciation for their efforts. As he supported them, they became enthusiastic supporters. His generals, well, that was another matter. Lincoln kept changing generals until he found Ulysses S. Grant who would fight and end the war, Lincoln's vision of the end game—to win. Grant's prewar experience and accomplishments paled in comparison to those of Robert E. Lee, but he was placed into a position by Lincoln where he could succeed.

Many of the issues facing society today require real leadership to create long-term solutions. The need to maintain the infrastructure that made our economy strong should be among those priorities—but a vision of future needs is also required. Think about the city of Los Angeles (L.A.). The only reason large numbers of people can live in L.A. can be traced to the existence of aqueducts that were built in the 1900s by William Mulholland under the guidance of Mayor Fred Eaton. Their vision was to grow L.A., but a serious limitation at that time was the water supply. The aqueducts were vital as the metropolis developed through the 1930s. So, in reality, leadership is defined by what is left behind, not by the current condition. It is accomplished by an adjustment in thoughts and actions in order to adapt to changing conditions.

A book that relates to this need and is especially apropos to engineers is *The Cult of the Mouse* by Henry Caroselli (2003), in which the author encourages creativity above profits in the workplace. Creativity is what will provide innovative solutions that change how people live. It also lends itself to an area where the patents and economic opportunities exist in the private sector. America rose to greatness in the twentieth century in large part because of automobiles, airplanes, and energy; these were the result of creative ideas that made many things possible.

If leadership skills were easily defined, there would be a lot more schools trying to provide them, and as a result, generations of leaders would be perpetually created. But these institutions do not seem to do well at teaching leadership. The public often looks to elected officials for leadership, but many pages could be written discussing the fallacy of that approach. No offense intended here, but can anyone really say that Millard Fillmore, Andrew Johnson, Franklin Pierce, James Buchanan, or Warren Harding were great leaders? They rank among *U.S. News*' worst 10 presidents of all time.

People see Lincoln very differently than these men, in part because he was able to lead us through difficult times. History treats Franklin D. Roosevelt, Teddy Roosevelt, and John F. Kennedy similarly. Likewise, we have recognized business leaders currently, but mostly these men and women are making money for their stockholders; few are making a big difference in society today. Yet, others such as Steve Jobs, Bill Gates, Paul Allen, Mark Zuckerberg, Henry Ford, and Thomas Edison are recognized because they have all made a difference in our lives and how we live, not because of their business acuity (Bloetscher, 2019). On the other hand, who is the CEO of Goldman Sachs? Most people won't know the answer to that question. That person is not considered a leader because the impact of his individual actions upon a normal person's life is basically indiscernible.

8.2 RESULTS OF AN ENGINEERING SURVEY ON LEADERSHIP

One of the goals of this survey is to query who those in the water industry identify (generally) as leaders, and determine, to some extent, where the leaders can be identified within these organizations.

These respondents would then be expected to identify barriers to leadership potential, as well as the means to create more leaders in the future. Comparing the perceptions of younger engineers to older practitioners was a key variable within the process. The survey was conducted using SurveyMonkey from June 2018 to October 2018, and it was freely available to anyone who had access to the survey link.

The survey incorporated a blog (publicutilitymangement.com) on WordPress® and several personal Facebook® pages that connected to over 2,000 persons. Two general email requests were also sent to a list of 6,000 people as part of the methodology within this endeavor.

The survey contained 41 questions regarding leadership, organizational details, personal experiences, demographics, and training. There were several major goals:

- To learn how professionals view their organizations and any barriers to leadership
- To assess training needs and current training providers
- To gain a particular understanding of the perceptions of various age groups—specifically the differences between those less than 35 years of age and those who are older

SurveyMonkey was used for data collection and analysis; Excel® was also used for data analysis. XLSTAT® was used for correlation analysis.

The respondents represented a total of 40 states, three Canadian provinces and a Federal District. Florida had the highest number of responses, followed by Georgia. Figure 8.1 shows the age ranges of the respondents. The respondents were overly represented by those more than 45 years old. More than half of the respondents had been in the industry for 20 years or more. No other experiential demographic received more than 15 percent (see Figure 8.2). More than 72 percent were male, and at least 80 percent were listed as Caucasian.

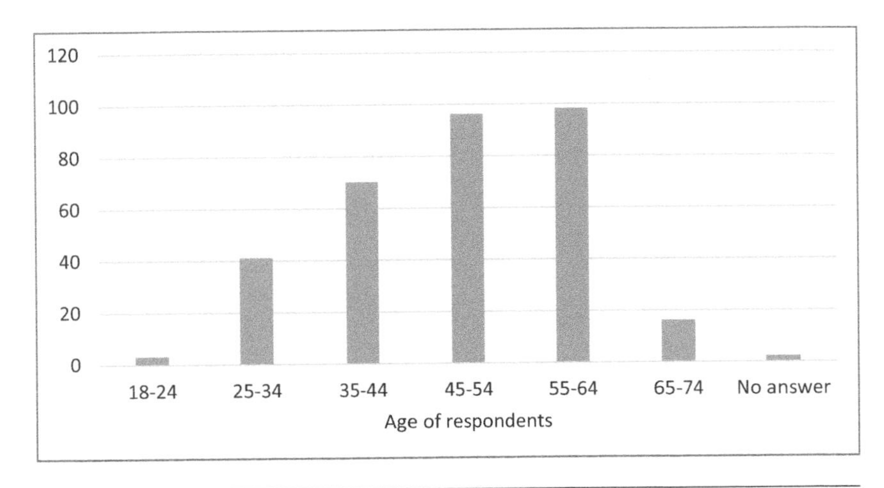

Figure 8.1 Age of respondents (n = 326)

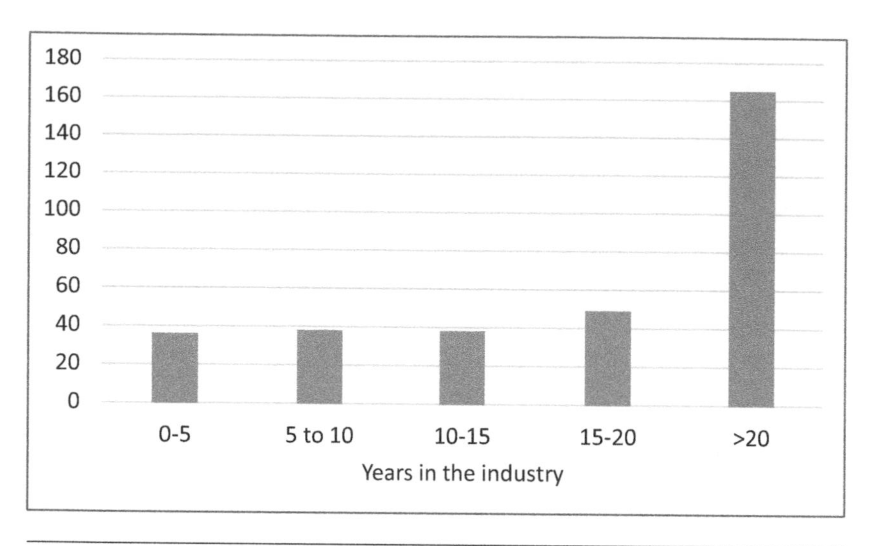

Figure 8.2 Years in the industry of respondents (n = 326)

More than 60 percent worked for utilities that were primarily large organizations. Of interest, many had not been with their current organization for an appreciable period of time, despite being within the industry for many years (see Figure 8.3). Nearly 75 percent had at least a four-year degree (see Figure 8.4). Ninety percent of the respondents were located in urban or suburban areas and more than 70 percent served within public agencies.

For comparison, this group is somewhat different than CEOs for the water industry. A paper by Teodoro and Whisenant (2013) indicated that CEOs were primarily male (93.9 percent), 50–60 years old (mean 53.8), white/non-Latino (95.8%), had been in their jobs for nearly 10 years, and reported having a BS degree (46%) in finance or science. Only 29% were educated as engineers. The 122 respondents served mainly large utilities, with a suggestion by the authors that smaller utilities had less access to people with comparable skill sets (Teodoro and Whisenant, 2013) (see Figure 8.5).

One of the survey questions asked respondents to identify challenges that they faced within their jobs. Too much politics, budgets,

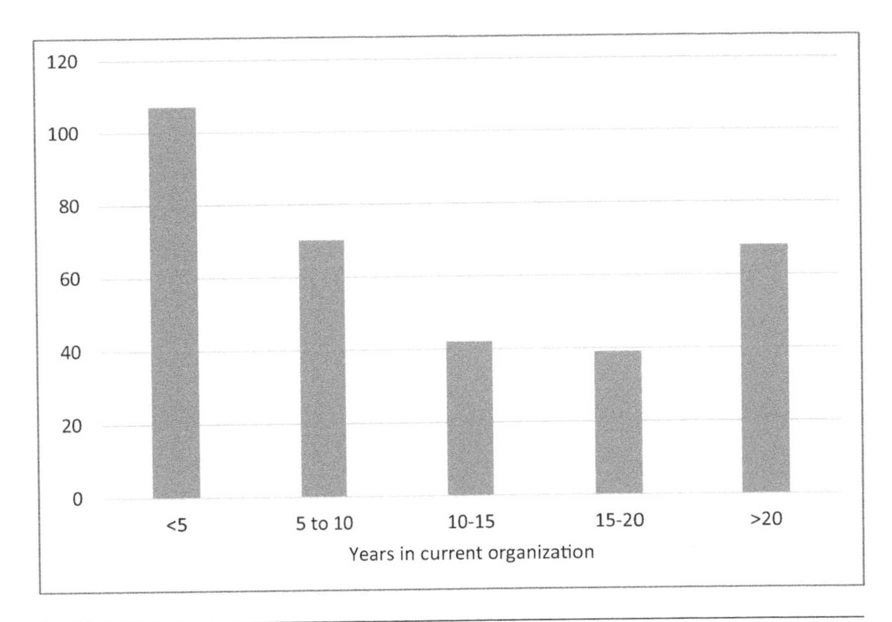

Figure 8.3 Years with current organization of respondents (n = 326)

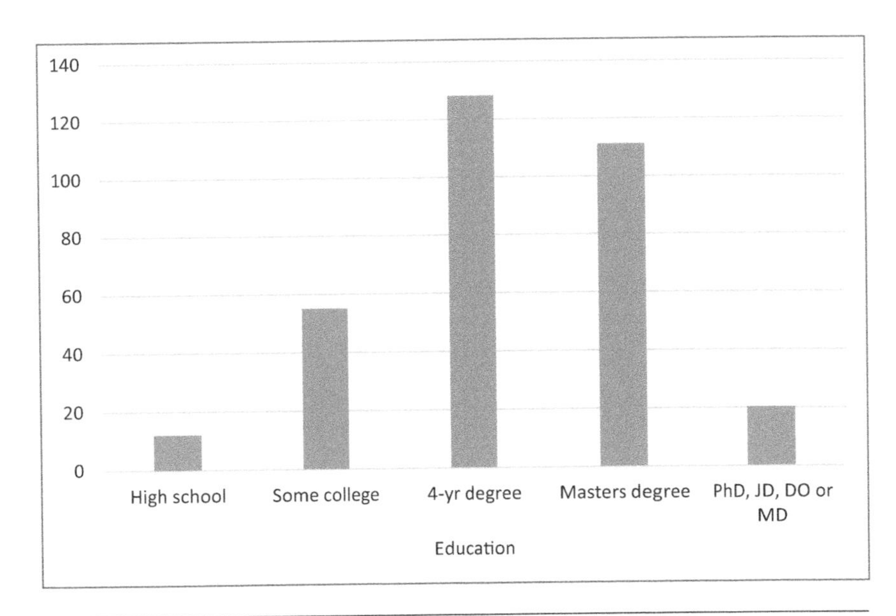

Figure 8.4 Education of respondents (n = 326)

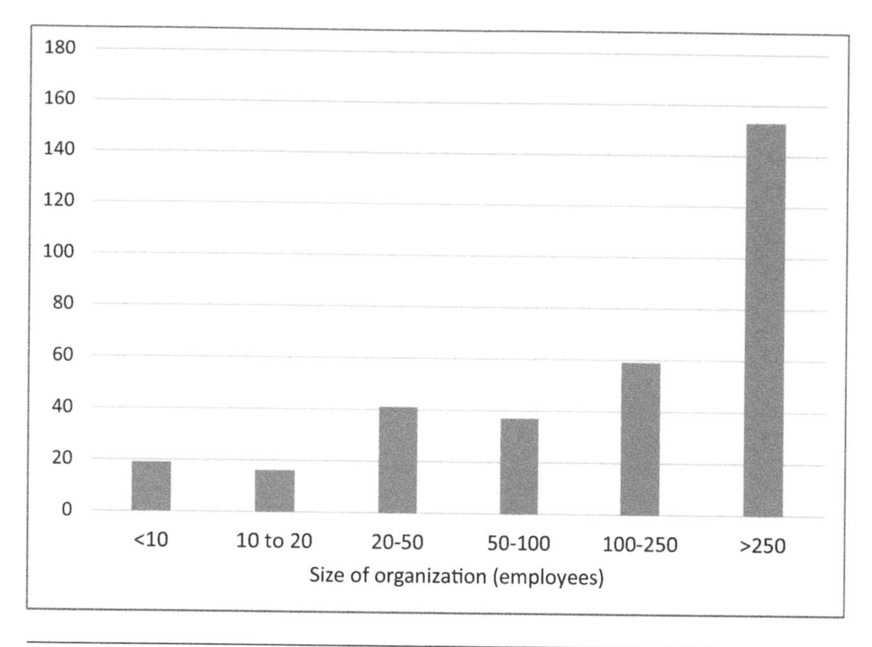

Figure 8.5 Size of organization of respondents (n = 326)

and lack of infrastructure to do their jobs were the major issues. Meanwhile, despite these challenges, the top priority for nearly two-thirds of the respondents was protecting public health.

Nearly 72 percent of respondents were able to perceive that leadership and management were defined separately, and that *building leadership for the future* was the most viable area in which to provide assistance to communities. Yet, when asked what they saw as barriers to the leadership that was currently being exhibited in their organization (see Table 8.1), it was noted that respondents indicated *organizational structure/norms*, followed by *a lack of opportunities* and *a lack of training* as most pressing. The results further revealed a perception that leadership was not demonstrated by policymakers in these organizations (20.74 percent).

Table 8.1 Barriers to leadership in the organization

Response	Percent Responding	Number of Respondents
Organizational structure/norms	51.1	165
Lack of opportunities	39.9	129
Training	33.8	109
Politics	27.9	90
Budget constraints do not allow the organization to exhibit leadership	20.7	67
Other (please specify)	15.2	49

The respondents were seeking leadership opportunities. Nearly half of the respondents had received more than 50 hours of leadership training, and much of that training was provided in-house or through training at a local university.

Most of the respondents believed that the training they received was useful, but identified the biggest leadership challenge in their job to be a lack of time to devote to improving themselves. The responses suggest that organizations were lean—perhaps too much so—and that there was insufficient time for planning, upgrades, and communication with the public. Since leadership rests within communication, planning, and setting a vision, overall, these respondents did not appear to believe that they had adequate time and resources to lead.

From where may future leaders emerge, and can they be expected to be materially different in their views and makeup than their older peers? For purposes of this section, young professionals (YPs) are identified as those in the industry who are 35 years of age or younger. This designation does create some limitations in the selection of a representative group of YP respondents, given that the YPs are relatively new to the profession and are less likely to have attended conferences or have their names included on mailing lists that might be connected to this survey.

With respect to race, Figure 8.6 shows that the YPs were more diverse than the non-YPs, but were still primarily caucasian. The YP respondents were 42 percent female versus 24.8 percent non-YPs. Of no surprise, the YPs have generally been in the industry less than 10 years, and within their organization for nearly five years. However, this encouraging indication may reflect that many may have come directly from college to the water industry, a good sign for the future.

The biggest issue identified by both groups was the existence and perception of too much politics. The YPs saw this as a bigger issue than the non-YPs. Interestingly, the YPs did not see infrastructure issues to be as critical as perceived by the older generation (see Figure 8.7).

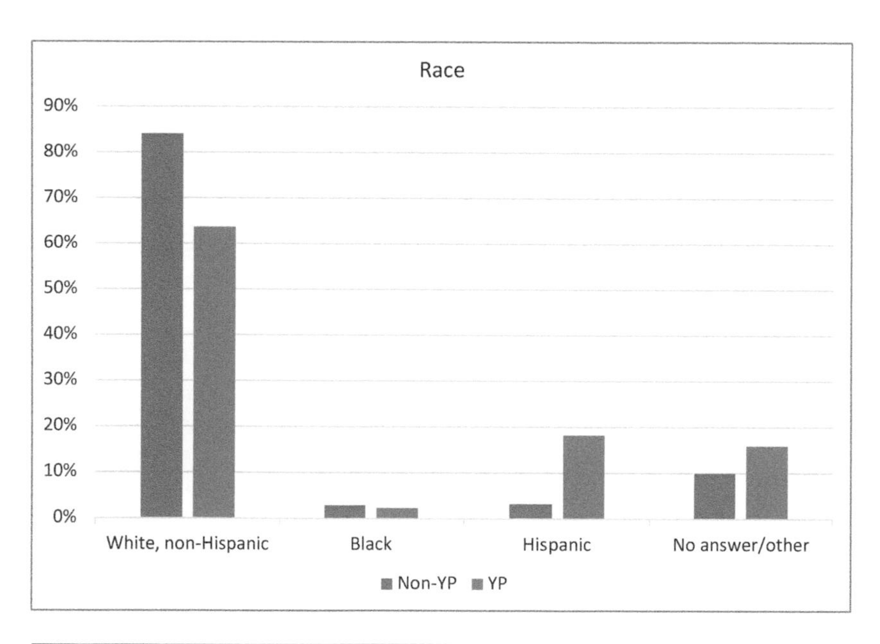

Figure 8.6 Race of YPs versus non-YPs (as a percent of respondents)

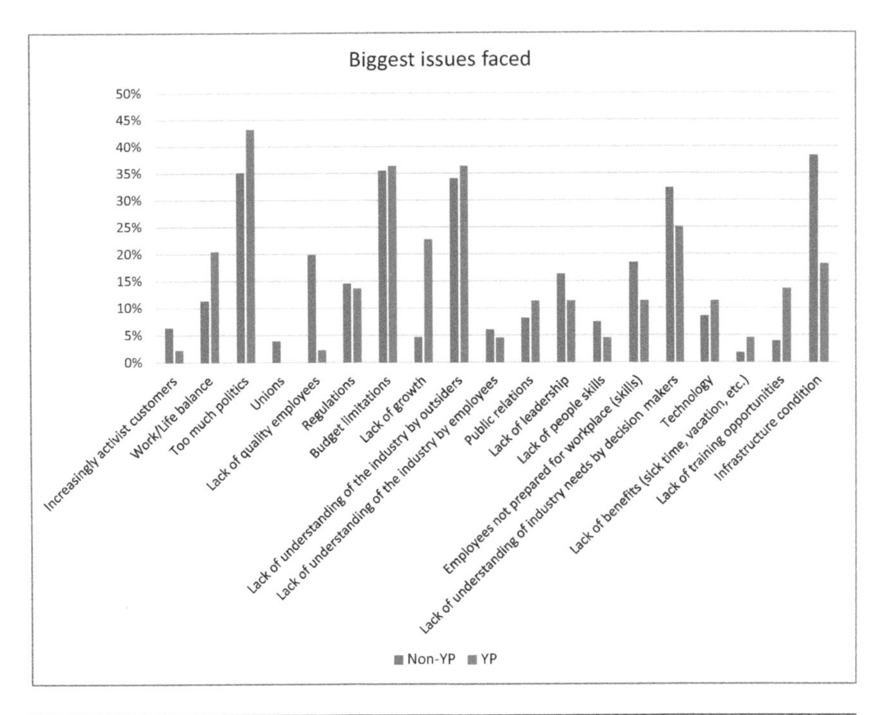

Figure 8.7 Biggest concerns or challenges faced by respondents

However, both groups had similar responses as to how the industry should provide leadership help to the community (see Figure 8.8). The YPs reported a lack of training opportunities and job opportunities as greater challenges with more frequency than the older respondents. Both groups agreed that the amount of time required to develop leadership skills was the greatest barrier to be faced (see Figure 8.9). YPs tend to watch the non-YPs and rely upon reading to understand leadership—i.e., the non-YPs are the role models for the YPs (see Figure 8.10).

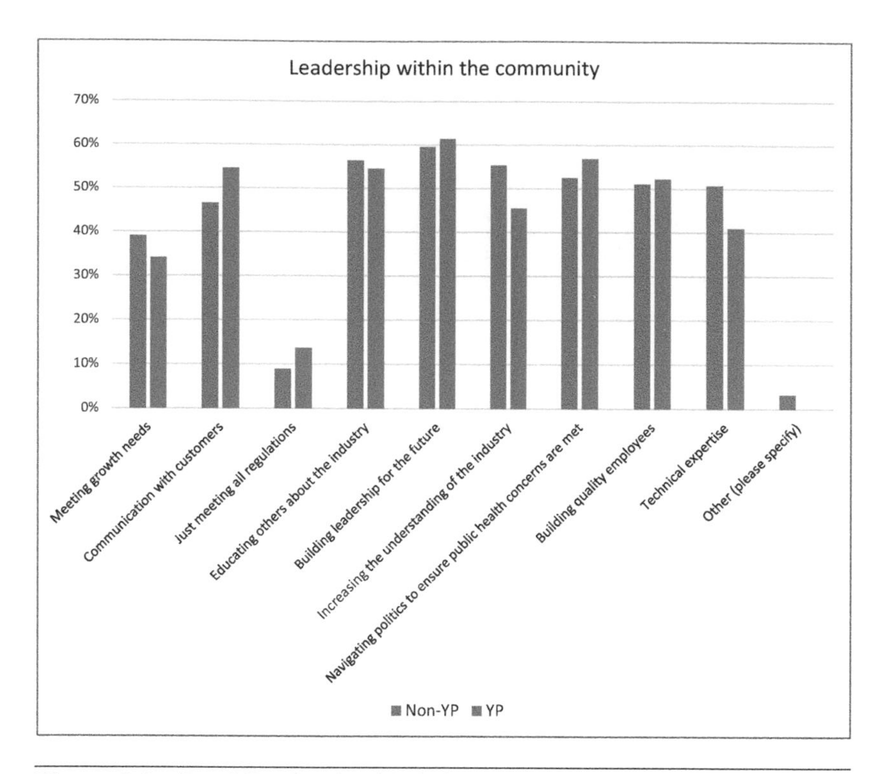

Figure 8.8 Providing leadership help

Figure 8.9 Barriers to leadership in the organization

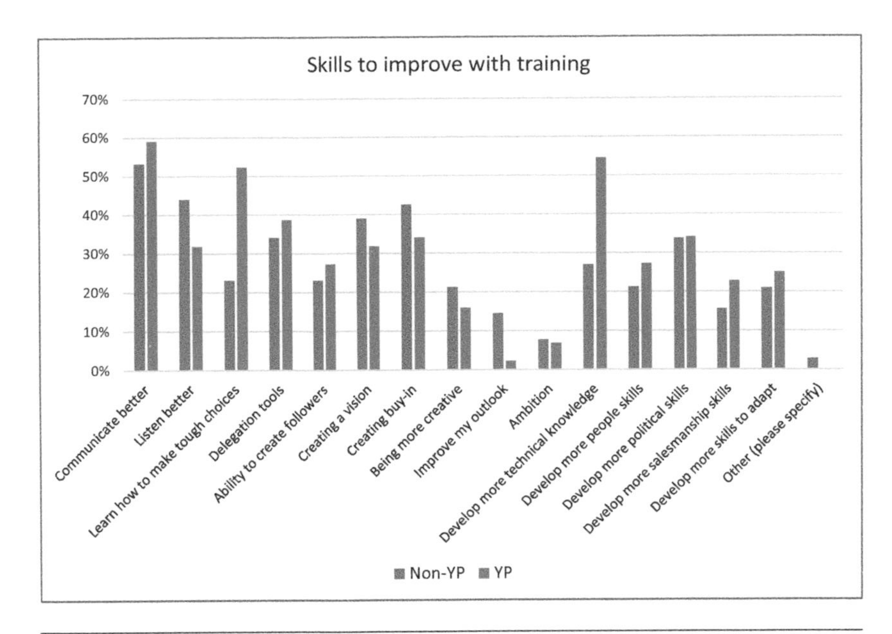

Figure 8.10 Skills desired to improve leadership potential

The YPs differ from non-YPs in other significant ways. YPs tend to identify leadership in those who deal in peace and bringing people together. They are far less likely to name coaches and business leaders as the source of leadership (see Figure 8.11). Those who inspire ranked second on the YP list, only behind leaders who bring people together (see Figure 8.12).

Finally, the YPs saw leadership within professionals more than non-YPs. No other category more greatly differed between groups. Neither group saw leadership at the executive level (see Figure 8.13).

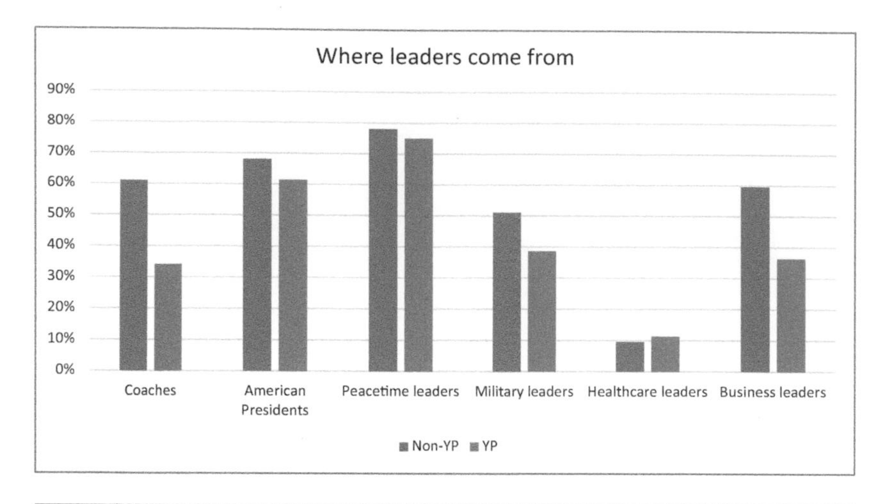

Figure 8.11 Where leaders come from in each group

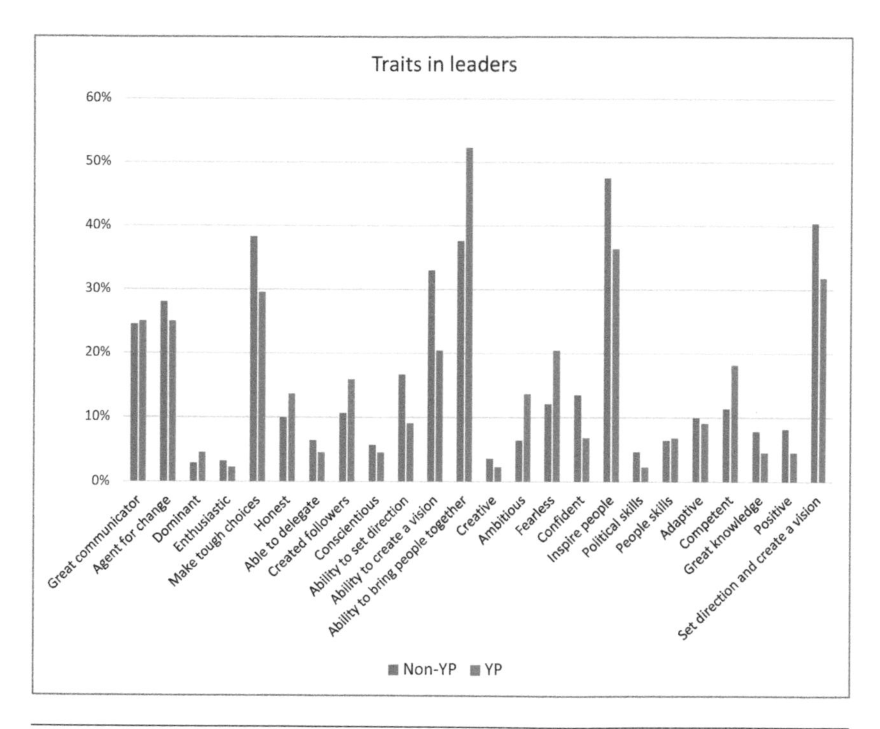

Figure 8.12 Skills found in great leaders

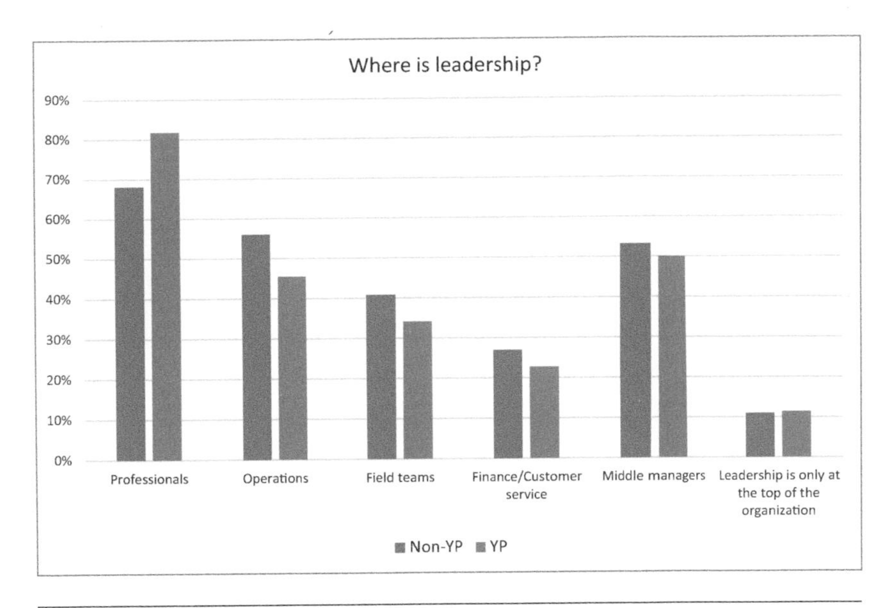

Figure 8.13 Where leadership exists in the organization

8.3 CONCLUSIONS

Leadership confronts all, and engineers are on the forefront of any challenge given their position as agents of social change. Engineers are leaders in that they change society with the products they design and build.

In a nutshell, society needs engineers—and engineers need society to have trust and confidence in their work. Ethics, education, and performance are the means to achieve this trust. As the survey indicates, professionalism is an important part of innovation and leadership. Leaders must emerge from the shadows and show society their potential to contribute to the welfare of all. That, in turn, will elevate engineers and create a better understanding of the impact of engineering on our economic and societal well-being. As shown in Figures 8.14–8.16, engineers build things that make a difference—both old and new.

Figure 8.14 Dam building on the Au Sable River that more than 100 years ago brought electricity to north central Michigan

Figure 8.15 Locks and canals at Sault Ste. Marie connect Lake Superior with Lake Huron. (Photo courtesy of Charles F. Steele.)

Figure 8.16 Construction of the Dania Beach nanofiltration water treatment plant was the first in the world to incorporate the concepts of *green building* to Gold Certification recognized throughout the world, proving that a utility plant could be both productive and green

REFERENCES

Bloetscher, F. (2019). *Public Infrastructure Management*. J. Ross Publishing. Plantation, FL.

Caroselli, H.M. (2003). *The Cult of the Mouse*. Ten Speed Press. Berkeley, CA.

This book has free material available for download from the Web Added Value™ resource center at *www.jrosspub.com*

CHAPTER 9

CONCLUSIONS

The engineering profession continues to evolve, and we can anticipate that it is likely to change dramatically over the coming decades. The computer skills that engineering graduates currently possess can be expected to be obsolete within 10 years. Computers make calculations faster, but the overdependence on programming creates a challenge—if engineers do not understand the methodology of how computer programs work to produce and resolve data input, they cannot have confidence in the deliverables that are being provided within their work assignments. Certain licensing boards now require engineers to thoroughly understand any software they may have to use.

As the engineering profession continues to encounter more complicated issues, the quantity of relevant academic credits that are required in academia continues to decrease. The competition from business colleges in the 1990s caused many engineering programs to decrease the number of credits that are necessary to graduate from anywhere between 140 and 145 to the current average of 130. In the process, important classes such as thermodynamics, circuits, chemistry, and additional physics and math were deleted within various programs. Currently, several state legislatures are pushing respective universities' systems programs to reduce the requirements to 120 credits for all degrees, including engineering.

Industry advisory boards at several of these schools suggest that decreasing credits may not be of benefit to the industry if technical classes are further reduced. This argument does not seem to sway the political powers. In contrast, the American Society of Civil Engineers had previously suggested 150 credits (basically a master's degree) be required for licensure, but this has failed to gain much traction among any state licensing boards.

The elimination of certain classes and the increased complexity within fields requires that virtually all engineering projects become multidisciplinary—long gone are the days when any individual engineer was tasked with the entire design. Different disciplines require greater knowledge of codes, materials, and operating parameters than any one individual can be expected to provide. The fast pace of changes within these engineering disciplines can be expected to create specific design impacts that will affect the capabilities of the individual as well.

Software fundamentally impacts the process of design and fabrication/construction. Schools need to keep up with evolving changes within artificial intelligence (AI), which reflects upon the skills of faculty members as well. However, capstone design classes are difficult to design as multidisciplinary, without cooperation and innovation by various members of any faculty. This type of approach has yet to bear the required involvement.

There are challenges to the concept of licensing. Over half of the states have introduced legislation during the past five years that is associated with at least one of two issues. One is a *right-to-engage-in-a-lawful-occupation* type of act that seeks to limit or lessen business or occupation regulation (NSPE, 2020). This legislation often proposes the elimination of licensure requirements for certain professions under the premise that a person has a *right to engage in a lawful profession or vocation without being subject to an occupational regulation.*

Many of these acts contain language that retains the regulatory authority on occupations that are necessary to meet public health, safety, and welfare objectives, requires review periods every four or five years, and is universally applicable to every licensing board.

Two states—Indiana and Missouri—introduced a *Consumer Choice* bill that would allow unlicensed persons to practice within occupations that require licensure, provided the persons disclosed that they are unlicensed (NSPE, 2020). Fortunately, both bills died in the legislature in 2019, but could return.

Neither bill was specifically oriented toward the engineering profession, but the legislation within the Indiana Job Creation Commission, that was introduced in 2014, would have recommended elimination of licensure of professional engineers (NSPE, 2020). The State of Montana introduced legislation in 2017 to eliminate licensing of professional engineers; this died in committee in 2017.

New Mexico's Governor issued an Executive Order (2018-048) which would have permitted unlicensed persons to perform work that otherwise would require licensure as long as the customer is informed and is willing to sign a contract acknowledging that the unlicensed person is doing such (NSPE, 2020). The Tennessee and West Virginia legislatures were also considering such a bill in 2020 (NSPE, 2020).

Sunset laws on licensing boards require a review and analysis of licensing boards (associated with any regulatory requirement that the respective board be dissolved) that typically results in recommendations to remove any unnecessary or overly burdensome licensing requirements, and the eventual recommendation for continuation of the board. Ohio enacted legislation that reads, "Ohio will use the least restrictive regulation to protect public health and safety" as a part of reviewing all licensing boards (NSPE, 2020).

None of these political moves would appear to meet the public health, safety, and welfare requirements that people expect of government. In addition, many states place restrictions on people who call themselves *engineers* if they do not have a license, purely to protect the public perception of the profession.

There will always be challenges as to how engineers do business. Technology can provide many advantages resulting in efficiency. The advancement of technology in the 1990s created major improvements within the productivity of the U.S. workforce. But the major change at that time was in the added use of machines at the expense of human personnel in manufacturing (Miller, 2017), which can be expected to continue into the foreseeable future (Conley, 2019).

It can be expected that improvements in technology will continue to impact middle income workers, who may require retraining to continue to participate in the workforce. Engineers should be cognizant of the unintended consequences of technological advancements and be ready to offer solutions.

Engineering the future also means re-engineering our society. In addition, concepts like AI may improve business decision making, but as Jamie Dimon, chairman and chief executive officer of JPMorgan Chase, noted on the television show *60 Minutes* in January 2020, improved decision making and business efficiency may disproportionately impact economic and socially vulnerable populations, thereby undoing years of effort to provide social equity for all. This is an inevitable truth.

Ethical issues impact procurement—lobbying elected officials and such—and would appear to contradict goals (and in some case statutes for public sector contracts) that suggest the best qualified engineers be chosen for jobs. Low bidding practices may still have gotten us to the moon, but was that the most efficient way to do so? Bidding on engineering contracts based on cost versus qualifications increases the risk to the client because the engineers involved may be less qualified and thus cheaper to hire. Qualification-based selection is intended to allow clients to acquire the expertise that would improve efficiency and lower risk to the client as well as the public at large.

One of the goals of this manuscript is to review how licensure and education can come together within the engineering profession to protect public health, safety, and welfare. For more than 100 years this intent has been met across the United States by teaching institutions and licensing boards. Both entities adapted along the way to new technology and more specialized fields of engineering.

However, the primary goals remain the same—to protect the public health, safety, and welfare. Often, young engineers are so involved with studies and work that they do not have the time to grasp these basic responsibilities. Required to understand more, these young engineers are challenged to grasp a basic public service orientation with responsibility to greater society within in a given curriculum. As an example, many schools still do not require their students to study ethical issues.

Certain challenges can be identified, most of which are driven by a political motive and sometimes seem to be in opposition to the protection of the public. Those external challenges can be concerning and bear great consideration for practicing engineers.

Not everyone has the capacity to be a good engineer. Bad engineers will invariably hurt the reputation of the profession, which is why disciplinary practices within all licensure exist. Efforts to curtail licensure are quite concerning—what engineers *do* has the potential for far greater harm than the risks that an individual doctor might encounter.

Compounding the degree of jeopardy that a licensed practitioner may face is the lack of public understanding of the degree of competence, efficiency, and forethought that is required of professionals within the field of engineering.

Engineers need to guard against any effort to minimize the skill set requirements that are needed to ensure an ethical approach to building our future. It is hoped that the public will soon routinely perceive and appreciate the differences between trained engineers and *train engineers.*

WHY THIS ALL MATTERS . . .

In closing out this text, a timely subject is the collapse of the 12-story Champlain Towers South condominium building in Surfside, FL, on June 24, 2021. Half a building collapsed on sleeping residents at 1:30 in the morning. Weeks were spent cleaning up the rubble and recovering bodies crushed under the concrete—a very sad and difficult endeavor. Over the next several years, you will read and hear a lot about this tragedy, and the finger-pointing and lawsuits will continue for decades. The Surfside collapse, however, was years in the making so here is a little background.

A condominium association is made up of the owners of units in a building. They elect a board who makes decisions about the building, which then often hires a management company to manage the day-to-day affairs. For the most part, the board is made up of people who know little about building infrastructure or the needs to maintain it. The condominium association owns and is responsible for the infrastructure (walls and floors) but not the interior property. The board may hire experts for many things like legal services, engineering, and accounting activities.

The Champlain Towers South condominium was 40 years old and built to a different building code than today. The condo building was on the beach and the underground garage appears to have been periodically inundated with salt water from both the ground and via saltwater tides. Also, the pool deck above was perfectly flat, so water literally went through the concrete into the garage. In 2018, the Surfside condo association hired an engineer who found significant concerns with respect to structural damage to the building. The cost estimate to fix the infrastructure was $15+ million dollars, which created an average assessment on the units of $120,000 (the units averaged $600,000 in value). Many residents were on fixed incomes so getting to a decision to assess this amount of money was a challenge. Many condos have found their whole boards replaced for proposing far smaller assessments. At the time of the collapse, it had been three years and none of the fixes had been done.

This type of trade-off between the necessary maintenance of deteriorating infrastructure and the funds needed to pay for it will be more and more common in the future. The infrastructure can be a building, bridge, sewer, water line, treatment facility, power grid, etc. The majority of our nation's infrastructure is in poor shape because we do not spend enough money to maintain the great works we have designed and built. Before 1980 we spent 3.4% of GDP on infrastructure. That number has been reduced to about 1.3% since, so it is no wonder our infrastructure went from a B to a D in 40 years based on the ASCE report card.

Protection of the public is part of our professional and ethical responsibility and is both a legal and a moral one. Thomas Hobbes believed in the concept of a social contract and our social contract is set by law: to protect the public health, safety, and welfare. As a profession, engineering is assumed to meet the tests of public interest, whereby engineers are focused on the *public good*. Just as the tragic Surfside collapse killed nearly 100 people, failure of our infrastructure risks lives. It is our job to do what we can to avoid future failures as we design new projects and maintain older ones. Let us use the loss in Surfside to rededicate ourselves to our higher public purpose.

REFERENCES

ABET. (2020). 2019–2020 Criteria for Accrediting Engineering Programs. https://www.abet.org/accreditation/accreditation-criteria/criteria-for-accrediting-engineering-programs-2019-2020/.

Connely, C. (2019). Robots May Replace 800 Million Workers by 2030. These Skills Will Keep You Employed. https://www.cnbc.com/2017/11/30/robots-may-replace-up-to-800-million-workers-by-2030.html.

Michigan Tech. (2020). 2020 Engineering Salary Statistics. https://www.mtu.edu/engineering/outreach/welcome/salary/.

Miller, C.C. (2017). *Evidence That Robots Are Winning the Race for American Jobs*. https://www.cnbc.com/2017/11/30/robots-may-replace-up-to-800-million-workers-by-2030.html.

Moore, G.E. (1965). Cramming More Components onto Integrated Circuits. *Electronics*. Retrieved 2/2/20.

NCEES. (2020). Squared. National Council of Examiners for Engineering and Surveying. https://ncees.org/wp-content/uploads/Squared-2019.pdf.

NSPE. (2020). Threats to Licensure State List. https://www.nspe.org/resources/issues-and-advocacy/action-issues/threats-professional-licensure.

APPENDIX A

Code of Ethics for Engineers

Preamble

Engineering is an important and learned profession. As members of this profession, engineers are expected to exhibit the highest standards of honesty and integrity. Engineering has a direct and vital impact on the quality of life for all people. Accordingly, the services provided by engineers require honesty, impartiality, fairness, and equity, and must be dedicated to the protection of the public health, safety, and welfare. Engineers must perform under a standard of professional behavior that requires adherence to the highest principles of ethical conduct.

I. Fundamental Canons

Engineers, in the fulfillment of their professional duties, shall:

1. Hold paramount the safety, health, and welfare of the public.
2. Perform services only in areas of their competence.
3. Issue public statements only in an objective and truthful manner.
4. Act for each employer or client as faithful agents or trustees.
5. Avoid deceptive acts.
6. Conduct themselves honorably, responsibly, ethically, and lawfully so as to enhance the honor, reputation, and usefulness of the profession.

1420 KING STREET, ALEXANDRIA, VIRGINIA 22314-2794 • 888-285-NSPE (6773) • LEGAL@NSPE.ORG • WWW.NSPE.ORG • PUBLICATION DATE AS REVISED JULY 2019 • PUBLICATION #1102

©NATIONAL SOCIETY OF PROFESSIONAL ENGINEERS. ALL RIGHTS RESERVED.

Code of Ethics for Engineers

II. Rules of Practice

1. **Engineers shall hold paramount the safety, health, and welfare of the public.**
   a. If engineers' judgment is overruled under circumstances that endanger life or property, they shall notify their employer or client and such other authority as may be appropriate.
   b. Engineers shall approve only those engineering documents that are in conformity with applicable standards.
   c. Engineers shall not reveal facts, data, or information without the prior consent of the client or employer except as authorized or required by law or this Code.
   d. Engineers shall not permit the use of their name or associate in business ventures with any person or firm that they believe is engaged in fraudulent or dishonest enterprise.
   e. Engineers shall not aid or abet the unlawful practice of engineering by a person or firm.
   f. Engineers having knowledge of any alleged violation of this Code shall report thereon to appropriate professional bodies and, when relevant, also to public authorities, and cooperate with the proper authorities in furnishing such information or assistance as may be required.

©NATIONAL SOCIETY OF PROFESSIONAL ENGINEERS. ALL RIGHTS RESERVED.

Code of Ethics for Engineers

2. **Engineers shall perform services only in the areas of their competence.**
    a. Engineers shall undertake assignments only when qualified by education or experience in the specific technical fields involved.
    b. Engineers shall not affix their signatures to any plans or documents dealing with subject matter in which they lack competence, nor to any plan or document not prepared under their direction and control.
    c. Engineers may accept assignments and assume responsibility for coordination of an entire project and sign and seal the engineering documents for the entire project, provided that each technical segment is signed and sealed only by the qualified engineers who prepared the segment.
3. **Engineers shall issue public statements only in an objective and truthful manner.**
    a. Engineers shall be objective and truthful in professional reports, statements, or testimony. They shall include all relevant and pertinent information in such reports, statements, or testimony, which should bear the date indicating when it was current.
    b. Engineers may express publicly technical opinions that are founded upon knowledge of the facts and competence in the subject matter.
    c. Engineers shall issue no statements, criticisms, or arguments on technical matters that are inspired or paid for by interested parties, unless they have prefaced their comments by explicitly identifying the

©NATIONAL SOCIETY OF PROFESSIONAL ENGINEERS. ALL RIGHTS RESERVED.

interested parties on whose behalf they are speaking, and by revealing the existence of any interest the engineers may have in the matters.

4. **Engineers shall act for each employer or client as faithful agents or trustees.**
    a. Engineers shall disclose all known or potential conflicts of interest that could influence or appear to influence their judgment or the quality of their services.
    b. Engineers shall not accept compensation, financial or otherwise, from more than one party for services on the same project, or for services pertaining to the same project, unless the circumstances are fully disclosed and agreed to by all interested parties.
    c. Engineers shall not solicit or accept financial or other valuable consideration, directly or indirectly, from outside agents in connection with the work for which they are responsible.
    d. Engineers in public service as members, advisors, or employees of a governmental or quasi-governmental body or department shall not participate in decisions with respect to services solicited or provided by them or their organizations in private or public engineering practice.
    e. Engineers shall not solicit or accept a contract from a governmental body on which a principal or officer of their organization serves as a member.
5. **Engineers shall avoid deceptive acts.**
    a. Engineers shall not falsify their qualifications or permit misrepresentation of their or their associates'

©NATIONAL SOCIETY OF PROFESSIONAL ENGINEERS. ALL RIGHTS RESERVED.

qualifications. They shall not misrepresent or exaggerate their responsibility in or for the subject matter of prior assignments. Brochures or other presentations incident to the solicitation of employment shall not misrepresent pertinent facts concerning employers, employees, associates, joint venturers, or past accomplishments.

b. Engineers shall not offer, give, solicit, or receive, either directly or indirectly, any contribution to influence the award of a contract by public authority, or which may be reasonably construed by the public as having the effect or intent of influencing the awarding of a contract. They shall not offer any gift or other valuable consideration in order to secure work. They shall not pay a commission, percentage, or brokerage fee in order to secure work, except to a bona fide employee or bona fide established commercial or marketing agencies retained by them.

III. Professional Obligations

1. **Engineers shall be guided in all their relations by the highest standards of honesty and integrity.**
   a. Engineers shall acknowledge their errors and shall not distort or alter the facts.
   b. Engineers shall advise their clients or employers when they believe a project will not be successful.
   c. Engineers shall not accept outside employment to the detriment of their regular work or interest. Before accepting any outside engineering employment, they will notify their employers.

©NATIONAL SOCIETY OF PROFESSIONAL ENGINEERS. ALL RIGHTS RESERVED.

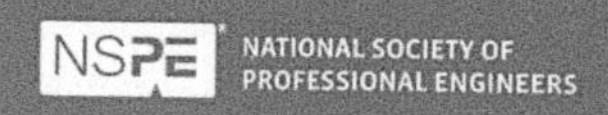

Code of Ethics for Engineers

d. Engineers shall not attempt to attract an engineer from another employer by false or misleading pretenses.

e. Engineers shall not promote their own interest at the expense of the dignity and integrity of the profession.

f. Engineers shall treat all persons with dignity, respect, fairness, and without discrimination.

2. Engineers shall at all times strive to serve the public interest.

a. Engineers are encouraged to participate in civic affairs; career guidance for youths; and work for the advancement of the safety, health, and well-being of their community.

b. Engineers shall not complete, sign, or seal plans and/or specifications that are not in conformity with applicable engineering standards. If the client or employer insists on such unprofessional conduct, they shall notify the proper authorities and withdraw from further service on the project.

c. Engineers are encouraged to extend public knowledge and appreciation of engineering and its achievements.

d. Engineers are encouraged to adhere to the principles of sustainable development[1] in order to protect the environment for future generations.

[1] "Sustainable development" is the challenge of meeting human needs for natural resources, industrial products, energy, food, transportation, shelter, and effective waste management while conserving and protecting environmental quality and the natural resource base essential for future development.

1420 KING STREET, ALEXANDRIA, VIRGINIA 22314-2794 • 888-285-NSPE (6773) • LEGAL@NSPE.ORG •
WWW.NSPE.ORG • PUBLICATION DATE AS REVISED JULY 2019 • PUBLICATION #1102
©NATIONAL SOCIETY OF PROFESSIONAL ENGINEERS. ALL RIGHTS RESERVED.

Code of Ethics for Engineers

e. Engineers shall continue their professional development throughout their careers and should keep current in their specialty fields by engaging in professional practice, participating in continuing education courses, reading in the technical literature, and attending professional meetings and seminar.

3. **Engineers shall avoid all conduct or practice that deceives the public.**
    a. Engineers shall avoid the use of statements containing a material misrepresentation of fact or omitting a material fact.
    b. Consistent with the foregoing, engineers may advertise for recruitment of personnel.
    c. Consistent with the foregoing, engineers may prepare articles for the lay or technical press, but such articles shall not imply credit to the author for work performed by others.
4. **Engineers shall not disclose, without consent, confidential information concerning the business affairs or technical processes of any present or former client or employer, or public body on which they serve.**
    a. Engineers shall not, without the consent of all interested parties, promote or arrange for new employment or practice in connection with a specific project for which the engineer has gained particular and specialized knowledge.
    b. Engineers shall not, without the consent of all interested parties, participate in or represent an adversary interest in connection with a specific project or

©NATIONAL SOCIETY OF PROFESSIONAL ENGINEERS. ALL RIGHTS RESERVED.

Code of Ethics for Engineers

proceeding in which the engineer has gained particular specialized knowledge on behalf of a former client or employer.

5. **Engineers shall not be influenced in their professional duties by conflicting interests.**
    a. Engineers shall not accept financial or other considerations, including free engineering designs, from material or equipment suppliers for specifying their product.
    b. Engineers shall not accept commissions or allowances, directly or indirectly, from contractors or other parties dealing with clients or employers of the engineer in connection with work for which the engineer is responsible.
6. **Engineers shall not attempt to obtain employment or advancement or professional engagements by untruthfully criticizing other engineers, or by other improper or questionable methods.**
    a. Engineers shall not request, propose, or accept a commission on a contingent basis under circumstances in which their judgment may be compromised.
    b. Engineers in salaried positions shall accept part-time engineering work only to the extent consistent with policies of the employer and in accordance with ethical considerations.
    c. Engineers shall not, without consent, use equipment, supplies, laboratory, or office facilities of an employer to carry on outside private practice.

©NATIONAL SOCIETY OF PROFESSIONAL ENGINEERS. ALL RIGHTS RESERVED.

Code of Ethics for Engineers

7. **Engineers shall not attempt to injure, maliciously or falsely, directly or indirectly, the professional reputation, prospects, practice, or employment of other engineers. Engineers who believe others are guilty of unethical or illegal practice shall present such information to the proper authority for action.**
    a. Engineers in private practice shall not review the work of another engineer for the same client, except with the knowledge of such engineer, or unless the connection of such engineer with the work has been terminated.
    b. Engineers in governmental, industrial, or educational employ are entitled to review and evaluate the work of other engineers when so required by their employment duties.
    c. Engineers in sales or industrial employ are entitled to make engineering comparisons of represented products with products of other suppliers.
8. **Engineers shall accept personal responsibility for their professional activities, provided, however, that engineers may seek indemnification for services arising out of their practice for other than gross negligence, where the engineer's interests cannot otherwise be protected.**
    a. Engineers shall conform with state registration laws in the practice of engineering.
    b. Engineers shall not use association with a nonengineer, a corporation, or partnership as a "cloak" for unethical acts.

©NATIONAL SOCIETY OF PROFESSIONAL ENGINEERS. ALL RIGHTS RESERVED.

Code of Ethics for Engineers

9. **Engineers shall give credit for engineering work to those to whom credit is due, and will recognize the proprietary interests of others.**
    a. Engineers shall, whenever possible, name the person or persons who may be individually responsible for designs, inventions, writings, or other accomplishments.
    b. Engineers using designs supplied by a client recognize that the designs remain the property of the client and may not be duplicated by the engineer for others without express permission.
    c. Engineers, before undertaking work for others in connection with which the engineer may make improvements, plans, designs, inventions, or other records that may justify copyrights or patents, should enter into a positive agreement regarding ownership.
    d. Engineers' designs, data, records, and notes referring exclusively to an employer's work are the employer's property. The employer should indemnify the engineer for use of the information for any purpose other than the original purpose.

"By order of the United States District Court for the District of Columbia, former Section 11(c) of the NSPE Code of Ethics prohibiting competitive bidding, and all policy statements, opinions, rulings or other guidelines interpreting its scope, have been rescinded as unlawfully interfering with the legal right of engineers, protected under the antitrust laws, to provide price information to prospective clients;

©NATIONAL SOCIETY OF PROFESSIONAL ENGINEERS. ALL RIGHTS RESERVED.

accordingly, nothing contained in the NSPE Code of Ethics, policy statements, opinions, rulings or other guidelines prohibits the submission of price quotations or competitive bids for engineering services at any time or in any amount."

Statement by NSPE Executive Committee

In order to correct misunderstandings which have been indicated in some instances since the issuance of the Supreme Court decision and the entry of the Final Judgment, it is noted that in its decision of April 25, 1978, the Supreme Court of the United States declared: "The Sherman Act does not require competitive bidding."

It is further noted that as made clear in the Supreme Court decision:

1. Engineers and firms may individually refuse to bid for engineering services.
2. Clients are not required to seek bids for engineering services.
3. Federal, state, and local laws governing procedures to procure engineering services are not affected, and remain in full force and effect.
4. State societies and local chapters are free to actively and aggressively seek legislation for professional selection and negotiation procedures by public agencies.
5. State registration board rules of professional conduct, including rules prohibiting competitive bidding for engineering services, are not affected and remain in full force and effect. State registration boards with authority to adopt rules of professional conduct may adopt rules governing procedures to obtain engineering services.

©NATIONAL SOCIETY OF PROFESSIONAL ENGINEERS. ALL RIGHTS RESERVED.

Code of Ethics for Engineers

6. As noted by the Supreme Court, "nothing in the judgment prevents NSPE and its members from attempting to influence governmental action . . ."

Note: In regard to the question of application of the Code to corporations vis-a-vis real persons, business form or type should not negate nor influence conformance of individuals to the Code. The Code deals with professional services, which services must be performed by real persons. Real persons in turn establish and implement policies within business structures. The Code is clearly written to apply to the Engineer, and it is incumbent on members of NSPE to endeavor to live up to its provisions. This applies to all pertinent sections of the Code.

1420 KING STREET, ALEXANDRIA, VIRGINIA 22314-2794 • 888-285-NSPE (6773) • LEGAL@NSPE.ORG •
WWW.NSPE.ORG • PUBLICATION DATE AS REVISED JULY 2019 • PUBLICATION #1102
©NATIONAL SOCIETY OF PROFESSIONAL ENGINEERS. ALL RIGHTS RESERVED.

APPENDIX B

Code of Ethics[1]

Fundamental Principles[2]

Engineers uphold and advance the integrity, honor and dignity of the engineering profession by:

1. using their knowledge and skill for the enhancement of human welfare and the environment;
2. being honest and impartial and serving with fidelity the public, their employers and clients;
3. striving to increase the competence and prestige of the engineering profession; and
4. supporting the professional and technical societies of their disciplines.

Fundamental Canons

1. Engineers shall hold paramount the safety, health and welfare of the public and shall strive to comply with the principles of sustainable development[3] in the performance of their professional duties.
2. Engineers shall perform services only in areas of their competence.
3. Engineers shall issue public statements only in an objective and truthful manner.
4. Engineers shall act in professional matters for each employer or client as faithful agents or trustees, and shall avoid conflicts of interest.
5. Engineers shall build their professional reputation on the merit of their services and shall not compete unfairly with others.
6. Engineers shall act in such a manner as to uphold and enhance the honor, integrity, and dignity of the engineering profession and shall act with zero-tolerance for bribery, fraud, and corruption.
7. Engineers shall continue their professional development throughout their careers, and shall provide opportunities for the professional development of those engineers under their supervision.
8. Engineers shall, in all matters related to their profession, treat all persons fairly and encourage equitable participation without regard to gender or gender identity, race, national origin, ethnicity, religion, age, sexual orientation, disability, political affiliation, or family, marital, or economic status.

Guidelines to Practice Under the Fundamental Canons of Ethics

Canon 1.

Engineers shall hold paramount the safety, health and welfare of the public and shall strive to comply with the principles of sustainable development in the performance of their professional duties.

a. Engineers shall recognize that the lives, safety, health and welfare of the general public are dependent upon engineering judgments, decisions and practices incorporated into structures, machines, products, processes and devices.
b. Engineers shall approve or seal only those design documents, reviewed or prepared by them, which are determined to be safe for public health and welfare in conformity with accepted engineering standards.
c. Engineers whose professional judgment is overruled under circumstances where the safety, health and welfare of the public are endangered, or the principles of sustainable development ignored, shall inform their clients or employers of the possible consequences.
d. Engineers who have knowledge or reason to believe that another person or firm may be in violation of any of the provisions of Canon 1 shall present such information to the proper authority in writing and shall cooperate with the proper authority in furnishing such further information or assistance as may be required.

With permission from ASCE

e. Engineers should seek opportunities to be of constructive service in civic affairs and work for the advancement of the safety, health and well-being of their communities, and the protection of the environment through the practice of sustainable development.
f. Engineers should be committed to improving the environment by adherence to the principles of sustainable development so as to enhance the quality of life of the general public.

Canon 2.

Engineers shall perform services only in areas of their competence.

a. Engineers shall undertake to perform engineering assignments only when qualified by education or experience in the technical field of engineering involved.
b. Engineers may accept an assignment requiring education or experience outside of their own fields of competence, provided their services are restricted to those phases of the project in which they are qualified. All other phases of such project shall be performed by qualified associates, consultants, or employees.
c. Engineers shall not affix their signatures or seals to any engineering plan or document dealing with subject matter in which they lack competence by virtue of education or experience or to any such plan or document not reviewed or prepared under their supervisory control.

Canon 3.

Engineers shall issue public statements only in an objective and truthful manner.

a. Engineers should endeavor to extend the public knowledge of engineering and sustainable development, and shall not participate in the dissemination of untrue, unfair or exaggerated statements regarding engineering.
b. Engineers shall be objective and truthful in professional reports, statements, or testimony. They shall include all relevant and pertinent information in such reports, statements, or testimony.
c. Engineers, when serving as expert witnesses, shall express an engineering opinion only when it is founded upon adequate knowledge of the facts, upon a background of technical competence, and upon honest conviction.
d. Engineers shall issue no statements, criticisms, or arguments on engineering matters which are inspired or paid for by interested parties, unless they indicate on whose behalf the statements are made.
e. Engineers shall be dignified and modest in explaining their work and merit, and will avoid any act tending to promote their own interests at the expense of the integrity, honor and dignity of the profession.

Canon 4.

Engineers shall act in professional matters for each employer or client as faithful agents or trustees, and shall avoid conflicts of interest.

a. Engineers shall avoid all known or potential conflicts of interest with their employers or clients and shall promptly inform their employers or clients of any business association, interests, or circumstances which could influence their judgment or the quality of their services.
b. Engineers shall not accept compensation from more than one party for services on the same project, or for services pertaining to the same project, unless the circumstances are fully disclosed to and agreed to, by all interested parties.
c. Engineers shall not solicit or accept gratuities, directly or indirectly, from contractors, their agents, or other parties dealing with their clients or employers in connection with work for which they are responsible.
d. Engineers in public service as members, advisors, or employees of a governmental body or department shall not participate in considerations or actions with respect to services solicited or provided by them or their organization in private or public engineering practice.
e. Engineers shall advise their employers or clients when, as a result of their studies, they believe a project will not be successful.

f. Engineers shall not use confidential information coming to them in the course of their assignments as a means of making personal profit if such action is adverse to the interests of their clients, employers or the public.
g. Engineers shall not accept professional employment outside of their regular work or interest without the knowledge of their employers.

Canon 5.

Engineers shall build their professional reputation on the merit of their services and shall not compete unfairly with others.

a. Engineers shall not give, solicit or receive either directly or indirectly, any political contribution, gratuity, or unlawful consideration in order to secure work, exclusive of securing salaried positions through employment agencies.
b. Engineers should negotiate contracts for professional services fairly and on the basis of demonstrated competence and qualifications for the type of professional service required.
c. Engineers may request, propose or accept professional commissions on a contingent basis only under circumstances in which their professional judgments would not be compromised.
d. Engineers shall not falsify or permit misrepresentation of their academic or professional qualifications or experience.
e. Engineers shall give proper credit for engineering work to those to whom credit is due, and shall recognize the proprietary interests of others. Whenever possible, they shall name the person or persons who may be responsible for designs, inventions, writings or other accomplishments.
f. Engineers may advertise professional services in a way that does not contain misleading language or is in any other manner derogatory to the dignity of the profession. Examples of permissible advertising are as follows:
   - Professional cards in recognized, dignified publications, and listings in rosters or directories published by responsible organizations, provided that the cards or listings are consistent in size and content and are in a section of the publication regularly devoted to such professional cards.
   - Brochures which factually describe experience, facilities, personnel and capacity to render service, providing they are not misleading with respect to the engineer's participation in projects described.
   - Display advertising in recognized dignified business and professional publications, providing it is factual and is not misleading with respect to the engineer's extent of participation in projects described.
   - A statement of the engineers' names or the name of the firm and statement of the type of service posted on projects for which they render services.
   - Preparation or authorization of descriptive articles for the lay or technical press, which are factual and dignified. Such articles shall not imply anything more than direct participation in the project described.
   - Permission by engineers for their names to be used in commercial advertisements, such as may be published by contractors, material suppliers, etc., only by means of a modest, dignified notation acknowledging the engineers' participation in the project described. Such permission shall not include public endorsement of proprietary products.
g. Engineers shall not maliciously or falsely, directly or indirectly, injure the professional reputation, prospects, practice or employment of another engineer or indiscriminately criticize another's work.
h. Engineers shall not use equipment, supplies, laboratory or office facilities of their employers to carry on outside private practice without the consent of their employers.

Canon 6.

Engineers shall act in such a manner as to uphold and enhance the honor, integrity, and dignity of the engineering profession and shall act with zero-tolerance for bribery, fraud, and corruption.

a. Engineers shall not knowingly engage in business or professional practices of a fraudulent, dishonest or unethical nature.
b. Engineers shall be scrupulously honest in their control and spending of monies, and promote effective use of resources through open, honest and impartial service with fidelity to the public, employers, associates and clients.
c. Engineers shall act with zero-tolerance for bribery, fraud, and corruption in all engineering or construction activities in which they are engaged.
d. Engineers should be especially vigilant to maintain appropriate ethical behavior where payments of gratuities or bribes are institutionalized practices.
e. Engineers should strive for transparency in the procurement and execution of projects. Transparency includes disclosure of names, addresses, purposes, and fees or commissions paid for all agents facilitating projects.
f. Engineers should encourage the use of certifications specifying zero-tolerance for bribery, fraud, and corruption in all contracts.

Canon 7.

Engineers shall continue their professional development throughout their careers, and shall provide opportunities for the professional development of those engineers under their supervision.

a. Engineers should keep current in their specialty fields by engaging in professional practice, participating in continuing education courses, reading in the technical literature, and attending professional meetings and seminars.
b. Engineers should encourage their engineering employees to become registered at the earliest possible date.
c. Engineers should encourage engineering employees to attend and present papers at professional and technical society meetings.
d. Engineers shall uphold the principle of mutually satisfying relationships between employers and employees with respect to terms of employment including professional grade descriptions, salary ranges, and fringe benefits.

Canon 8.

Engineers shall, in all matters related to their profession, treat all persons fairly and encourage equitable participation without regard to gender or gender identity, race, national origin, ethnicity, religion, age, sexual orientation, disability, political affiliation, or family, marital, or economic status.

a. Engineers shall conduct themselves in a manner in which all persons are treated with dignity, respect, and fairness.
b. Engineers shall not engage in discrimination or harassment in connection with their professional activities.
c. Engineers shall consider the diversity of the community, and shall endeavor in good faith to include diverse perspectives, in the planning and performance of their professional services.

---

[1]The Society's Code of Ethics was adopted on September 2, 1914 and was most recently amended on July 29, 2017. Pursuant to the Society's Bylaws, it is the duty of every Society member to report promptly to the Committee on Professional Conduct any observed violation of the Code of Ethics.
[2]In April 1975, the ASCE Board of Direction adopted the fundamental principles of the Code of Ethics of Engineers as accepted by the Accreditation Board for Engineering and Technology, Inc. (ABET).
[3]In October 2009, the ASCE Board of Direction adopted the following definition of Sustainable Development: "Sustainable Development is the process of applying natural, human, and economic resources to enhance the safety, welfare, and quality of life for all of the society while maintaining the availability of the remaining natural resources."

INDEX

Page numbers followed by "*f*" refer to figures and those followed by "*t*" refer to tables.